D0154744

FORENSIC DIGITAL IMAGING
AND PHOTOGRAPHY

FORENSIC DIGITAL IMAGING AND PHOTOGRAPHY

Herbert L. Blitzer and Jack Jacobia

ACADEMIC PRESS
A Division of Harcourt, Inc.

San Diego   San Francisco   New York   Boston
London   Sydney   Tokyo

363.25
B649f
2002
V.1

This book is printed on acid-free paper.

Copyright © 2002 by ACADEMIC PRESS

All Rights Reserved
No part of this publication may be reproduced or transmitted in any form or by any means, electronic or mechanical, including photocopying, recording, or any information storage and retrieval system, without permission in writing from the publisher.

ACADEMIC PRESS
A Division of Harcourt Inc.
Harcourt Place, 32 Jamestown Road, London NW1 7BY, UK
http://www.academicpress.com

ACADEMIC PRESS
A Division of Harcourt Inc.
525 B Street, Suite 1900, San Diego, California 92101-4495, USA
http://www.academicpress.com

ISBN 0-12-106411-5
ISBN CD-ROM 0-12-106412-3

Library of Congress Catalog Number 2001092786
A catalogue record for this book is available from the British Library

Typeset by Kenneth Burnley, Wirral, Cheshire, England
Printed in Spain by Grafos SA Arte Sobre Papel, Barcelona
02 03 04 05 06 07 GF 9 8 7 6 5 4 3 2 1

CONTENTS

METHODIST COLLEGE LIBRARY
FAYETTEVILLE, NC

FOREWORD

Law enforcement agencies have long recognized the value of photography in the investigation and prosecution of criminal cases. The regular use of photographic images by police began in Paris in 1841 when photographs of criminals were taken as a means of establishing a rogues' gallery – the first primitive mugshot database. Since then, photography has become such an intrinsic part of law enforcement activities that it is almost impossible to imagine an investigation that did not involve photography in some way.

Photographs and videotapes recorded at crime scenes help investigators and juries understand spatial relationships between objects and people in the scene that would be difficult, if not impossible, to explain with mere words. So, too, do photographs taken of victims help document the nature of their wounds and injuries. The value of these images is often magnified by the fact that small details within an image that might not appear important when first encountered may later be discovered to be crucial to the investigation. By capturing details in a photograph they are not only preserved, but their characteristics may also be examined more carefully and in a way that permits them to be placed within the context of the broader investigation. Images help investigators "see the big picture."

Not only are the images of crime scenes or victims after the fact crucial to successful investigations, but so, too, are images captured either intentionally by security cameras, or inadvertently by eyewitnesses, victims, or criminals themselves. One need only consider the images taken by a surveillance camera showing the Ryder truck approaching the Murrah Federal Building in Oklahoma City in April 1995 or the photographs showing O. J. Simpson wearing Bruno Magli shoes in Buffalo's Rich Stadium to appreciate the value of such images – they bear silent witness quite relevant to later events.

Forensic laboratories are responsible for the generation of many images utilized by law enforcement today. Sometimes such images document the condition of evidence before and after an examination is conducted on the item – such as when a swatch of bloody cloth is removed from a piece of clothing as part of a DNA analysis. At other times, the images document the findings of a forensic examination – such as when a hair found at a crime scene is shown to be similar to that of a suspect through a side-by-side microscopic comparison or when the lands and grooves on a bullet recovered from a scene are shown to match those of a test bullet fired from a suspect's firearm.

Many of the photographs that are acquired in a field setting or in the laboratory will later be subjected to image processing or further scientific analysis as a means of extracting additional information from the image content. For example, a poorly illuminated scene depicted in a video image may be enhanced in order to improve the visibility of features in the shadow areas, or the image of a bloody fingerprint left on a piece of fabric may be processed to reduce the prominence of the weave of the fabric, permitting a successful comparison of the fingerprint.

Although these and other imaging activities have been utilized in law enforcement for decades, the past few years have witnessed an incredible expansion in the nature of the technologies available to conduct those investigations. In particular, the conversion from analog imaging technologies, such as silver-based film and analog video, to purely digital technologies, has been occurring at an almost overwhelming rate. This conversion to digital imaging systems raises the hopes and concerns of many law enforcement imaging professionals. This text addresses many of those hopes and concerns.

Digital imaging systems offer many benefits not immediately available with film or traditional analog video. Among these benefits include the ability to instantly review an image once it has been taken and the ability to easily import an image into digital image processing applications, where the image can be enhanced to improve the visibility of details in the image. Another major benefit of digital imaging is the ease with which images can be filed, stored, and transferred between locations and investigators. Digital imaging also offers an easy means of building image databases that have a variety of applications in law enforcement, such as mug shots or gang tattoos.

On the other hand, the advent of digital imaging in law enforcement is not without its serious concerns, as well. Among the many concerns regarding digital imaging include worries that digital images are easier to manipulate than film images and are not as trustworthy. There are also valid concerns that the image quality available through digital systems cannot match or even approach that of film. Finally, and a most critical issue for those concerned with the quality of images acquired by law enforcement professionals, is the misperception that digital imaging must be better than traditional film imaging because it is easy enough for anyone to do. Unfortunately, making something easy to do does not ensure that it will be done well.

This concern cannot be dismissed lightly, particularly when the images are utilized in a forensic setting – a setting in which the innocence or guilt of an individual is usually at question and mistakes could lead to either imprisonment for the innocent or freedom for the guilty. The individuals purchasing and using photographic equipment in a criminal justice environment must be aware of the fact that the acquisition of high quality images depends on far more than the camera used to record the images. In order to create the most detailed rendering of a scene with sufficient information to be useful to investigators and forensic scientists, the photographer must make decisions regarding such factors as proper lighting, depth of field, and color balance, to name but a few. Properly selecting and correcting for these factors cannot be programmed into a camera – these adjustments must be made by the photographer for every image that is taken. Every scene is different, and different items of evidence may require different procedures. As with any other technical field, to achieve a high level of competency and proficiency in forensic imaging requires extensive training, skill, and experience.

It was because of these concerns that the Scientific Working Group on Imaging Technologies (SWGIT) was formed. The mission of SWGIT is ". . . to facilitate the integration of imaging technologies and systems within the criminal justice system (CJS) by providing definitions and recommendations for the capture, storage, processing, analysis, transmission, and output of images"* ("SWGIT – Definitions and Guidelines for the Use of Imaging Technologies in the Criminal Justice System", *Forensic Science Communications*, July 2001). Fortunately for SWGIT and the law enforcement imaging community as a whole, Herb Blitzer and Jack Jacobia and their colleagues at the Institute for Forensic Imaging have prepared this text – *Forensic Digital Imaging* – which addresses many of our common concerns in a timely fashion. Not only are the most important issues and caveats related to digital imaging discussed in this book in a clear and understandable manner, but real-life solutions are offered that will permit law enforcement agencies to assure that their imaging practices are on a sound technical footing that will stand up in court proceedings.

The SWGIT has identified the following categories of users in the criminal justice system who should have a high degree of awareness and/or skills in imaging technologies: managers, supervisors/commanders, officers, crime scene technicians, criminalists/examiners, forensic photographers/videographers, lawyers, judges, legal assistants, and trainers/educators. Individuals in each of these categories need to understand the nature of the evidence they are dealing with, whether they are responsible for funding the acquisition of imaging equipment or whether they are making legal rulings on the admissibility of imaging evidence in a courtroom. An understanding of these issues is offered in this text in a way that will benefit members of every category of user defined above, as well as others who may utilize imaging in their careers, such as insurance adjusters.

For those responsible for interpreting the content of an image, it is necessary to understand how the human visual system works and the limitations imposed by the photographic process – whether it is a traditional silver-based process or one that relies on electronic processes. For those responsible for decisions regarding the purchasing of photographic equipment, an understanding of the capabilities of different kinds of equipment is essential. In today's environment, anyone in law enforcement considering moving from silver-based to digital imaging would be well advised to assess the quality of their current photographic products to insure that when they make a change they are still capable of meeting their mission-specific needs. If one has decided to begin working with digital images, it is important to understand how the pixels in the image can be changed individually or as a group through processing activities, including compression. It is especially critical to understand the difference between what *can* be done digitally and what *should* be done in a forensic setting.

The reader of this text will find useful information on all of these topics and more. Even more importantly, the law enforcement professional who follows the recommendations provided in this book can feel confident that he or she will be handling their imaging evidence in a manner that will stand up to the high standards necessary for the presentation of evidence in criminal cases.

In closing, let me offer two stories of bank robbers that hold lessons for the law enforcement imaging professional. In the first case, three bank robbers enter a bank – two tall guys and a short guy. After announcing their intentions, the two tall guys pick up the short guy and toss him over the counter, where he grabs the money and the three depart. Three weeks later, the same three robbers burst into the same bank and repeat the robbery exactly as it happened before. In another three weeks it happens again. After this third robbery the bank manager gets fed up and installs bullet-proof glass over the teller counter. Like clockwork, the three bank robbers arrive again to make their "withdrawal," only to be faced with the new scenario. The two tall guys take one look at the glass, look at each other, shrug, pick up the short guy and hurl him into the glass. He bounces off. They try again. Same result. As they approach the third guy for another try, he shakes his head and refuses. They argue. They continue to argue until the police arrive and arrest them.

In the second story, a lone robber enters a bank brandishing a pistol and quickly moves behind the counter where two tellers are working. He yells at both of them to put money into his bag. He watches as they try to comply. One of the tellers, however, does not work as rapidly as the robber would like. He focuses his attention on her and gets so frustrated with her inability to put the money in the bag fast enough that he decides to help her out. He places his pistol on the counter between himself and the second teller and moves to the first teller. During this activity, the second teller carefully picks up the pistol, places it into an open drawer, and closes the drawer. The robber finishes collecting the money from the first teller, collects the money from the second teller and starts to leave. He stops, however, realizing that he has forgotten something, but he can't quite place his finger on it. Then it hits him – WHERE'S MY GUN? He searches madly for the gun until he hears the police sirens, responding to the silent alarm triggered shortly after his entrance. Finally, realizing that his freedom is at stake, he dashes out the door and into the street, where he is promptly run down by the police car responding to the call. He does not survive the accident.

The first case illustrates the importance of planning, adaptability, and flexibility in the performance of one's job. The three robbers knew of only one way to perform their job. When the conditions changed, they did not have another plan, nor were they adaptable enough to figure out another way to successfully achieve their goal. They could have tried running around the counter. They might also have considered casing the bank before the fourth attempt to ensure that the same conditions existed. Law enforcement imaging professionals do not usually get to plan out the conditions in which they work. We must deal with the crime scenes and evidence that come to us in the best way we can. Good planning is helpful, but the ability to analyze a new environment and plan a different approach is a must.

New technologies, such as digital imaging, can help the law enforcement imaging professional be more adaptable since they offer so many different ways in which to handle imaging evidence. However, the second bank robber story should not be forgotten in the context of new technology. In this case, the bank robber became so focused on his pistol – his technology – that he forgot the main purpose of his activity – to get the money from the bank and then escape with his freedom. The pistol was merely a means to the end of getting the money. In much the same way, law enforcement imaging professionals must always remember that their imaging technologies, whether traditional silver-based film, analog video, or digital imaging, are simply a means of acquiring and processing evidence in the interest of seeing that justice is done. It is not the technology that is important, but the results of using that technology that matter. By following the guidance offered in this text, you'll not only handle imaging technologies better, but the ultimate results of your work will be better, too.

RICHARD VORDER BRUEGGE, PHD

ACKNOWLEDGEMENTS

Some have estimated that it took some 100,000 slaves to build the Great Pyramid at Giza. More recently, it would appear that in reality 20,000 willing workers did the job. Although the logic here is stretched, the 5 to 1 gain in productivity is of the right order of magnitude. We were blessed with willing workers and we wish to recognize some of them here.

First there were those that helped make the Institute for Forensic Imaging a reality. This grouping included IUPUI Chancellor Gerald Bepko, and Deans Alan Potvin, Doris Merritt, and H. Öner Yurtseven. In addition there were key members of the IUPUI Faculty, including Irv Levy, Russell Eberhart, Barry Bullard and Mohammed El Sharkaway. From the community we received support from Senator Richard Lugar, Governor Evan Bayh, Mayor Stephen Goldsmith, Chief Justice Randall Sheppard and Marion County Prosecuting Attorney Scott Newman. We also received the support of Leonard Redon, General Manager and Vice President of the Government Markets Division of Eastman Kodak and Ted Endo, Vice President of Canon USA, and Richard McEvoy.

The Institute has also enjoyed the gracious support of its board of directors, which includes Charles Burch, of the Indiana Law Enforcement Academy, Indiana State Police Major Robert Conley, Theodore Englehart, Michael Gentile, Amy Leitch Esq., Richard Kammen, Esq., and Alan Kimball.

There were also many who have helped us with mastery of content. This grouping includes IUPUI faculty members: Crystal Garcia, Jeffrey Huang, William Lin, Marvin Needler, pathology drs, Edward Berbari, Jose Adflk, and Greg Smith. From various criminal justice agencies we enjoyed the assistance of Poiu San Fillippo, Robert Warshaw, Donald Warshaw, John Mann, Ronald Blacklock, Diane Tolliver, Diane Donnelly, David Wadffg, Richard Vorder-Breugge, Michael Baden, George Reise, James Hamby, Sonia Leerkamp, Esq., James Cleek, Jon Daggy and Jay Barlow.

In addition there were several members of the IUPUI staff, including Kari Blackley, Sheri Alexander, and Nancy Ciskowski. And, the key IFI staff members were the people that made major contributions to the actual writing of the book: Research Scientist Suzali Suyut, Assistant Laboratory Director Leigha Hedrick, Director of Training Programs Jack Rubak, and Office Manager Kristy Trumpey.

Finally, this book is dedicated to the many students who have attended IFI courses and encouraged us to undertake this challenge.

CHAPTER 1

INTRODUCTION

Photographic systems are designed to substitute a constructed image in place of a natural scene. That is, the photograph is supposed to bring to the viewer a facsimile of what would have been seen had the viewer been at the original setting. Accordingly the system must respond to the original setting in a way that can render an output that the viewer will recognize as the original. This implies that the response of the photographic system must anticipate the human visual system both in how it captures the original image and how it produces the output image. The clear implication is that a basic knowledge of the human visual system is a prerequisite to developing and, to some degree using, a photographic system.

SEEING IMAGES

The human visual process is extremely complicated in that it is interactive with the scene being viewed. Its interactive nature makes it a remarkably powerful system: capable of very high resolution, very large dynamic range, very high sensitivity, and ability to see color and motion. A full description of this process is beyond the purview of this book, but there are a few important features of the human visual system that are important to understanding the field of inanimate imaging.

First and foremost, it is important to keep in mind that people see in their brains, and that the eye is merely a device that aids in the process. The eye takes in information, partially processes it, passes it to the optic nerve, which processes it some more, and then finally passes it to the brain where it is interpreted, or "seen." The various processing stations along that pathway are able to combine and store some information, and also able to send commands back to the eye. The result is that the visual system allows one to see more than one could see if the eye were a passive device, like a camera.

The best way to start to understand the visual system is with a description of the basic structures of the eye. Human color vision starts with the light-sensitive portion of the eye, known as the retina. This holds a very large number of individual receptors that act as the light-sensitive elements. They are very small, tubular structures arrayed as if standing, closely packed on a surface. Their active surfaces are all arranged to receive incoming light. Some of these receptors respond to a very broad range of colors and have a very high sensitivity. These have shapes that are very nearly proper cylinders and are called rods. They are the primary sensors for seeing in dark settings. They can respond to very low levels of light – as few as 10 or so photons are all that is required (the photon is the basic, individual particle of light, and light beams comprise large numbers of these traveling in a uniform direction). Rods are not sensitive to any particular color, however, so the brain does not ascribe color to the images derived from rod-based vision. The nighttime world is primarily black and shades of gray.

But there is another type of receptor, the cone. Again the name is based upon shape. Each of the cones is sensitive to a limited portion of the spectrum and they come in three basic varieties – sensitive to blue, green, and red light. The actual color sensitivities are rather peculiar compared to what one might imagine, but, clearly, they work just fine (see Figure 1.1). The cones have a lower general level of sensitivity than the rods and they are the basis for viewing in normal daylight. This is referred to as photopic vision. The rods and cones actually move relative to each other in response to the lighting level in the scene. In the dark the rods are extended relative to the cones, and in brighter light, the cones are extended. As we all know, it takes a few seconds for the eye to adapt to rapid and dramatic changes in light levels. Each of the light-sensing elements is connected through intermediary cells to so-called ganglion cells in

Figure 1.1

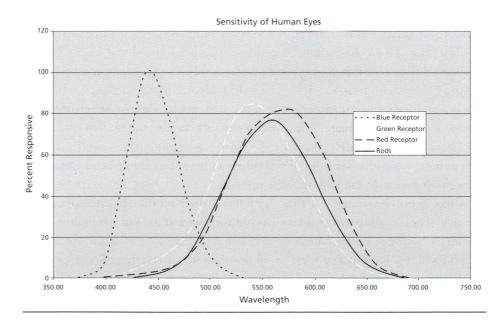

the retina. These cells form a complex network of local interconnections and are the means for starting the process of converting a set of individual light readings into that which the brain will conceive of as a color image.

Both rods and cones are activated by light falling on a receptor fluid that is pumped into them. Once the fluid has been fully used, it is replenished by fresh material. This process takes time, and gives rise to a phenomenon known as persistence of vision. Basically, the rods and cones respond and the system holds the readings until the fluid can be replenished. This can happen faster in bright light than in the darkness, and the result is that sensitivity to motion is not as good in the dark. Also, since only the rods are active in the dark, image sharpness is compromised. Thus it is hard to read fine print in weak light. This replenishment process also causes delays in adaptation to new light levels. Persistence of vision effects can be seen by staring at a bright, colored object for some period of time, and then rapidly shifting to looking at a neutral, uniform, less bright surface. If the original object was bright green, the viewer will see a magenta after-image appearing on the neutral surface. The after-image will fade away in a short time. Video and movie systems take advantage of this phenomenon by presenting a series of still images, each slightly different than the last in a systematic way. The individual frames are presented in a succession that is more rapid than the persistence of vision. The result is that the brain does not see a series of stills, but an image with apparent motion.

Getting back to the retina, close neighbor receptor cells with different color sensitivities allow the determination of red, green, and blue light falling at that location when their signals are combined. As mentioned earlier, the rods are sensitive to intensity only – not to color. They have a sensitivity peak, which is close to that of the green cones. As it turns out, there are many more green cones than either red or blue, and there are more red cones than blue. The result is that the eye has relatively good sensitivity to fine detail in the green portion of the spectrum, and somewhat less resolving power in the other two colors. The general lesson is that the eye has a tendency to see shape detail in the image based upon information in the green portion of the spectrum and then add color to this image. The green record also carries the bulk of the information regarding overall level of illumination. The result is that a photographic system must have relatively high discrimination for green information, and can afford to have less in the red and blue portions.

It has been clearly proven that light has a dual nature. It can appear as packets of energy, or particles known as photons, and at the same time it is built of electromagnetic waves. These waves comprise electric and magnetic fields, which oscillate regularly as the packet moves forwards. In Figure 1.2, the electric field is shown and the length of each wave is shown. Wavelength is directly related to the energy of the packet and also affects the color that the brain will

Figure 1.2

Electric wave field schematic

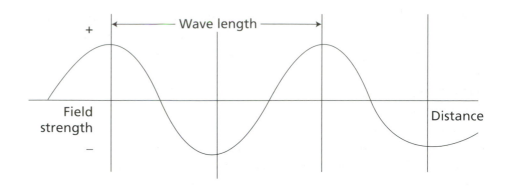

ascribe when it senses such packets. Longer wavelengths are less energetic and are associated with the color red, mid-length wavelengths have more energy than the red, and are associated with green, and the shorter wavelengths have more energy yet and are seen as blue. Physicists normally think of color in terms of the wavelength of the associated light. Blue colors are generally said to fall in the range 400–500 nm. Greens fall in the range 500–600 nm, and reds fall in the range 600–700 nm. Some of the other colors can be found at specific locations along this spectrum (Figure 1.3). For example, in the range immediately adjacent to 575 nm, the light appears to be yellow. And at the really short wavelengths, light appears violet.

But this wavelength-based process is not adequate for dealing with visual color. The eye has some unique capabilities and limitations. First of all, the main limitation – the eye cannot see outside the range 400–800 nm, the so-called visible spectrum. Light with a wavelength shorter than 400 nm is called ultraviolet, and there are photographic systems that can work in this band. Above 800 nm is the infrared spectrum, and there are photographic approaches that work well at "seeing" this light. The unique capability of key importance is that of the visual system, unlike photographic systems, to attach the subjective impression of color. The eye senses the wavelength by determining the extent to which it excites the different sensitivity cones. The brain then attaches the

Figure 1.3

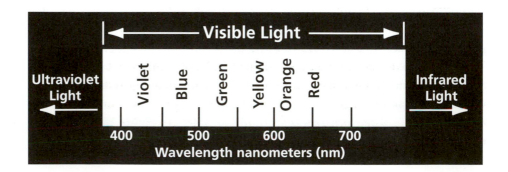

subjective notion of color. If there are combinations of light with different wavelengths, two or even all three cone types will be activated and the visual system will combine the inputs and impose a color. For example, where there are several broad, but uneven mixtures of light, the brain will see "brown." There is no wavelength associated with brown. It is a creation of the brain in response to the signals delivered by the visual system.

There can be some problems with this capability, though. If one were looking at an object that reflected a lot of light in the portion of the spectrum surrounding 575 nm, and not reflecting elsewhere, the object would appear to be yellow. Now note what will happen if one were to look at an object that reflected green light and red light in the portions of the spectrum close to 550 nm and 650 nm, respectively. Further, assume that there was no light at 575 nm (the portion that was just "seen" as yellow) or other portions of the spectrum. Nonetheless, the brain would consider this object to be yellow. Thus two objects, with very different color reflection profiles would give the same visual response. There are many such situations. Two objects' colors that appear the same when seen under one type of illuminant, will very likely appear to be different shades when seen under different lighting. These are referred to as metameric pairs. As we all know, there are many so-called optical illusions, in which the visual system can be "fooled." It is not our purpose here to list them all, but several of the color vision issues are important because many forensic applications are greatly aided by color photography and the photographer must allow for the differences between what is seen at the original setting and what shows up later in the photograph. The simple example is the fact that an at-the-scene observer will see that scene the same under both fluorescent and tungsten illumination, but unless a photographic system is properly adjusted, the resulting images will be seen as different. The eye/brain system adjusts for the setting and the camera may not unless the photographer takes specific steps to adjust it.

In practice, objects have a characteristic reflective profile, or spectral reflectance. They reflect certain amounts of light at each of the various wavelengths across the full spectrum. The three different types of cones (red, green, and blue) in the retina are stimulated accordingly and register how much light they each receive. The information is eventually passed to the brain where the combination of responses is combined and a color is ascribed to each particular portion of the scene. This is referred to as the tri-stimulus model of color vision. This model holds that combinations of three basic, or primary, colors can evoke all of the normal colors that can be seen. Thus any imaging system in which each element of an image can be represented by three values, representing the amounts of the three primary colors, will be able to provide an acceptable replica of the colors in an original object.

The visual primary colors are red, green, and blue, as described thus far. They are also referred to as the additive primary colors because they are used by adding one primary color of light to the other(s). Computer display screens and television sets are based upon these colors and utilize very small red, green, and blue lights to produce the color images we see on their surfaces. At normal viewing distances, the individual lights cannot be resolved as separate entities and they merge together such that their respective lights add together in the eye. To see this, select a yellow patch on your computer screen or TV screen and look closely at that spot with a magnifying glass. You will see spots or lines of red and green light in the area (the black you see is due to the shadow mask used to separate the color spots) – but there are no yellow spots! The yellow you see is in your head. It is simply that red light plus green light results in the perception of yellow. When all three primary colors, red, green, and blue, are added together in roughly equal proportions, the result is seen as white or gray, depending on total level. The absence of all of them is seen as black.

When red and blue are combined, the result is called magenta. When red and green are combined the result is yellow, and when green and blue are combined the color is referred to as cyan. These new colors, cyan, magenta, and yellow are called the subtractive primary colors. To understand why, consider that to create one of them, one uses two of the three additive primaries, but not the third. So, for example, magenta comprises red and blue – but no green! Therefore, magenta is the opposite, or negative of green. It can be thought of being created by taking white light and removing the green. Likewise, the opposite of blue is yellow and the opposite of red is cyan. Subtractive primaries are typically thought of as filters or dyes, while the additive primaries are lights. The subtractive primary filters each block one of the additive primary lights. If one were to superimpose cyan, magenta, and yellow filters and place the combination in a light beam, it would result in dark gray or black.

Printers utilize these subtractive primary colors to produce the image you see. If there is no ink, or dye, the paper is white. Add some cyan, and some red light is no longer reflected, but green and blue are unaffected. Likewise if yellow is applied, blue is no longer reflected but red and green are. And, add magenta and green is no longer reflected but blue and red are. Add magenta and yellow dyes together, and blue and green are blocked, but red comes through. And so it goes. Figure 1.4 shows the so-called color wheel. Think of three flashlights shining on a wall. One has a red filter over its face and the other two have green and blue filters respectively. Where all three beams superimpose, the patch is white. In the areas where pairs of lights superimpose, one of the subtractive primaries is formed. That subtractive primary will be on the opposite side of the diagram from its opposite, additive primary. So yellow is opposite blue, etc. In summary, a photographic system needs to have spatially distributed elements that are sensitive to red, green, and blue light (somewhat

Figure 1.4

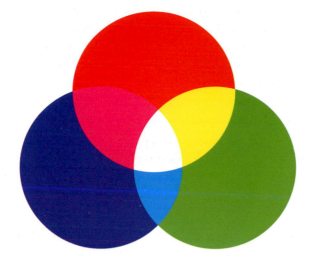

like the eye). And, it also needs the ability to attenuate light at specific locations using either adjustable red, green, and blue light sources (TV screen) or adjustable cyan, magenta, and yellow dyes (printed page). This will produce a system capable of producing color images that will be recognized as fair representations of original scenes.

Studies have shown that, while the eye responds to the three additive primary colors, the brain works with three other aspects of color. First of all there is the general level of illumination, sometimes referred to as value or luminance. Then there is the notion of a predominant color, referred to as hue (these are the subjective colors that the brain sees, not the pure wavelength colors). Finally, there is the notion of saturation. Highly saturated colors are very pure and can be quite vibrant. Desaturated colors seem as if they have been diluted with gray or white. One mapping of these dimensions is the CIE diagram (Figure 1.5). This is a solid geometric shape in which position along the vertical axis is indicative of level of illumination. The centerline of the core of this diagram represents the locus of all the grays, from black to white. These colors have value but no hue and no saturation. Flat planes radiating from the centerline towards a line on the periphery hold the colors of the same hue. Concentric cylindrical replicas of the periphery of the solid represent colors of equal saturation. Saturation increases as one moves from the centerline to the periphery. This, and related representations of colors, have been shown to be useful to designers of imaging systems because they are reasonable representations of how the eye sees different colors. One particular value of this type of representation is that one can change the level of saturation of a color without changing its hue or value. Or, one can adjust the hues without changing the overall level of brightness, etc. This is not true when one works with either the additive or

Figure 1.5

The CIE system (Commission International de L'eclairage or International Commission on Illumination), based on the concept of additive color mixing, is derived from experiments in which colors are matched by mixing colored lights. Though logical and straightforward, the arguments leading up to the rational basis of the CIE system are not the simplest to write down or to grasp.

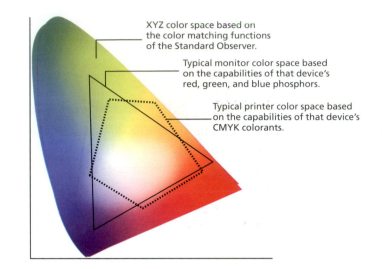

XYZ color space based on the color matching functions of the Standard Observer.

Typical monitor color space based on the capabilities of that device's red, green, and blue phosphors.

Typical printer color space based on the capabilities of that device's CMYK colorants.

subtractive primary colors. If one changes any one of the values of the primary colors, all three of the visual factors will change in a non-intuitive way. Representing colors as combinations of primary colors is essential for image capture and presentation, but difficult to use in image processing. More visually oriented systems are better for this purpose.

There is another color representation that is useful and is an extension of the three-color subtractive primary system. It is referred to as cyan, magenta, yellow, black, or CMYK. In the subtractive tri-color system, each location in an image is expressed as the amounts of cyan, magenta, and yellow needed to have the image seen as the right color. This color space is known as CMY. Note that equal amounts of cyan, magenta, and yellow dye will result in some level of gray, which can be thought of as a thin layer of black dye. It is possible to reduce equal amounts of cyan, magenta, and yellow dye and replace them with an equivalent amount of gray. This color space is particularly useful in dealing with matters relating to printers.

Still another way to represent colors is to estimate the overall level of illumination (as the rods and green cones might) and let this value carry a representation of green as well, then add information to adjust for redness and blueness. This approach has been particularly useful with electronic image-recording devices such as television, and other electronic cameras. There are mathematical formulae for converting from one color space to the other (although some information can be lost in the process), and when using the new technologies, it is helpful to be aware of the various systems. The overriding issue is that the human vision system, while it has some limitations, is able to adapt to various portions of a total scene and then combine several partial inputs to create an overall view. It also has the ability to identify objects in the scene, and ascribe features to these that might not otherwise be distinguishable (e.g., the object that recedes into the background and stays the same size even though the image of it formed by the eye's lens makes it continuously smaller). This same set of abilities allows humans to use pictures to stand as surrogates for real objects (most animals cannot do this).

As one looks at a scene, the eye does not see it as a fixed, whole scene. Instead the eye actually darts from one location to another in the scene, repeating important areas many times. It focuses on small portions of the full scene, or a set of objects, one at a time. The resulting image fragments are ultimately fused together by the brain to create a view of the original scene and all the objects therein. As each portion is viewed, there is feedback to the eye, which facilitates adjustments. For example, in bright portions, the eye adjusts to high-illumination viewing and can see detail in highlight portions. When the gaze is moved to a darker portion, the eye adjusts to dark-field viewing. And, when the elements are fused together in the brain, it will be possible to see detail in both the highlight and shadow portions of the scene "at the same time." In addition, the ganglion cells in the retina combine cone cell inputs in such a way as to accentuate transitions. The result is that the view has more resolving power than would otherwise be expected if the process were simply passive. In addition, the eye refocuses as it darts around the scene; so in the fused image, everything seems to be in focus all at once. Another adaptation is the binocular nature of human vision. The pair of views, one from each eye, are combined to determine how far away the various elements of the scene are from the viewer. This combination of capabilities allows the person to see a high resolution, over a wide field of view, and to see into both very bright and very dim portions of the image all at the same time – a remarkable feat!

But there is more. As was mentioned, there is a hierarchical visual system, but it does not stop with viewing alone. The brain is able to recognize features in the image and attach meaning to them very rapidly. The result is the ability to correct for distortions of objects. In passive systems, the mapping of three-dimensional objects onto a flat image results in distortions. For example, if one is looking at an object that extends from the close foreground to the background, such as a fallen telephone pole, it appears to be in focus over its entire length, it appears to be of consistent size, and does not appear to curve. In fact if there is a curve, or a reduction in diameter, the eye will probably be able to determine that because the pole changes at a rate inconsistent with other aspects of the total image. No current photographic system can do all of this, and when one considers that the visual system can do this all fast enough to play basketball, it truly humbles the finest team of imaging system engineers imaginable.

In summary, the human visual system really sees in the brain. Color is determined by combining inputs from three selectively colored receptors, so that photographic systems need only three properly selected colors to evoke all of the visible colors. The eye has a very great ability to deal with focus at all distances at once, to correct for perspective, and to see very high resolution. The result is that photographic systems will all fall short of the mark. This means that there is always room for improvement in photographic systems – that is to say, make the photographic systems better, and in all probability, the viewer will be able to appreciate the improvement.

While there are several different ways to represent the visible colors, each of them has two important requirements. First of all, they all define colors in terms of three numbers (CMYK uses four, but can easily be reduced to three). Secondly, there are mathematical formulae that allow translation from one space into any of the others by means of simple arithmetic. This means that scanners, and digital and video cameras can capture images in red, green, blue (RGB) terms, the images can be processed in CIE, or some other visually oriented space, and the images coming from the printer can be prepared using CMYK.

FILM IMAGING

There are three main components to this approach:

1. the sensitized materials upon which visual images are formed, that is, film or paper coated with a thin layer sensitized to light with silver halide (AgX) compounds,
2. devices used to project the images onto the materials, that is, the cameras and enlargers, and
3. processing equipment to render the images visible.

The camera is simply a device that uses a lens to create an image of a given scene at a point in space. It holds a piece of AgX film (or paper in the case of the enlarger) at the point where that image is formed, and when its shutter is opened, it projects the scene onto the film. It has control elements to help assure that a viable and useful image is created. In traditional photography, the camera can be used with a wide variety of films, lenses, and other accessories, making it a very versatile tool.

The sensitized material comprises a support sheet of film or paper that has several layers of light-sensitive material, the emulsion, carefully coated on to its surface. There may be more than a dozen separate layers in the emulsion, each only a few millionths of an inch thick. They are coated one on top of the other. In order to simplify this discussion, it will be comcerned with the major issues

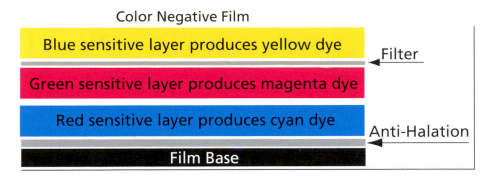

Figure 1.6

and the most common film – color negative. In these films there are a few layers near the top that are all sensitive to blue light. After processing, these layers produce yellow dye in the areas where blue light landed, and the amount of yellow dye will be increased in relation to the amount of blue light. Note that the film will be dark where there was originally more light, and the color will shift from the additive primary blue to its opposite, subtractive primary, yellow. In Figure 1.6, one can see the yellow dye in those topmost layers.

Just below the blue-sensitive layers, there is usually a filter layer that prevents blue light from penetrating any further. This is because AgX is intrinsically sensitive to blue light, and if it were to penetrate further, it would activate the other, lower layers as well. But these lower layers are supposed to respond to green and red respectively, and not respond to blue. The filter solves what would otherwise be a problem. The filtering material in this layer will be removed when the film is processed.

Below the filter layer there are typically a few layers that are sensitive to green light but not red light. These layers create a magenta dye when the film is processed. Again the material is designed so that more light results in more dye. Finally, near the bottom of the stack there are some layers that are sensitive to red light but not green. These layers produce cyan dye in response to exposure and processing.

At the very bottom, or even on the back of the film support, there is usually another filter layer that absorbs light broadly across the spectrum. This is called the anti-halation layer and it is there to prevent light from reflecting off the back of the film and creating secondary exposures on any return trips through the emulsion. Without this filter small rings tend to appear as little halos around bright objects.

As indicated above, in color negative, AgX film, the image that is produced is darker where the original subject was brighter, and the colors that are produced are the subtractive primaries that are opposite to the original additive primaries in the exposing beam of light; hence the reason for calling this a "negative" film, and the first image a negative. It is not intended for viewing; the final image will be created on paper that works in a very similar fashion, and thus makes a negative of a negative, or a positive image. Notice that the sensitive element in the negative film responds to the three additive primary colors and the print responds to these with the three subtractive primary colors such that every point in the image is represented by these three values. This fulfills the requirement for a viable photographic system. In addition there are two distinct steps to the process – making the negative and then making the (positive) print. In the first instance, the setting dictates many of the parameters employed in capturing the image (light level, coloration, brightness of the light objects relative to the dark ones, etc.). But, in the second, which is accomplished in the enlarger in the laboratory, the operator has the ability to control many of the parameters and settings. This affords real advantage, and over the years, photographers have developed a rich array of darkroom techniques that can be used to enhance and even modify images.

The light-sensitive material in all the active layers is AgX. This material is present in the form of very small crystals, like grains of table salt, only much smaller. They are suspended in a thin layer of gelatin. The layer also contains other materials, including: sensitizer dyes that make the crystals either more responsive to light in general (optical sensitizers), or sensitive to one of the primary colors (spectral sensitizers). In addition to these, there are large organic molecules called "colored couplers," which are basically dye molecules that are one color when the film is first coated and then are changed to another color when the film is processed – the conversion is in proportion to the original amount of crystal activation when the film is exposed. In the processed negatives, the colored coupler retains its original color in the unexposed areas. In areas of high levels of exposure, it will be largely converted to a dye that conforms to one of the subtractive primary colors. There will be gradations in between. The colored couplers are present to compensate for the unwanted absorption associated with the color dyes that constitute the actual image. Figure 1.7 shows typical absorptions for a common set of image-forming color dyes.

Remember that cyan dye is intended to modulate red light – and only red light. But as can be seen in the figure, the typical cyan dye also blocks a substantial amount of green light and blue light. Thus the dye that is formed in the red-sensitive layer is also affecting the passage of green and blue light, creating a problem. It is said to have unwanted absorption in the blue and the green. It behaves as if it were a combination of a lot of a pure cyan dye with a little yellow and magenta dye mixed in as well. In order to reduce the degree to which the dye in the red layer is affecting the green and blue layers the cyan colored coupler starts out as a combined yellow and magenta dye. This means that there is dye in the red layer that blocks a certain amount of blue and green light even

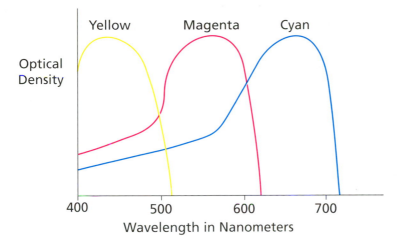

Figure 1.7
Spectral density of dyes.

if there is no red exposure. In areas where red exposure does occur, the coupler is converted to the cyan dye shown in the figure. This is the same dye that unfortunately happens to have a bit of magenta and yellow dye component as well. But since it both starts out with yellow and magenta, and ends up with yellow and magenta if it is converted, the impact on the blue and green records is balanced out. A similar effect occurs between the magenta dye and the blue portion of the spectrum. This result is that the coupler in the green-sensitive layer starts out as yellow. It only changes to magenta as dictated by exposure and processing. Thus the impact on the blue light band is evened out also. The overall result of the red-sensitive layer coupler starting out as yellow and magenta, and the green-sensitive layer coupler starting out as yellow, means that the film will show a salmon color combination of magenta and yellow one sees when looking at a processed negative in the areas that did not receive exposure. Slide films cannot be made with colored couplers; so making print images from negatives will result in inherently better results than making prints or duplicates from slides. Today's film scanners anticipate this problem, and it is possible to make corrections for the unwanted absorptions using the application software supplied. So scanners can provide high-quality images from either negatives or slides.

It was mentioned above that there are some advantages to negative systems over slides. One key advantage is that negative films can have considerably more exposure latitude, or dynamic range. The eye, with its ability to adapt locally to light levels is able to see both dark shadow detail and bright highlight detail in the same scene. One can easily see both the white-on-white finery of the bride's white-on-white gown and the black-on-black tuxedo on the groom standing at her side, or see the detail in the ice on the ground at the same time as seeing the small black animal hiding under a dark log. This ability to deal with both darks and lights at the same time is what is meant by dynamic range. The white on white can be up to a million times brighter than the black on black and a human can still see it all – seemingly all at once. No photographic system is so tolerant. While the human visual system can deal with ranges in the area of one million to one, color negative films are more likely to be in the range of 20,000 to 1 or so. The term exposure latitude is sometimes used in this context. If a film does not have much exposure latitude, the camera must be set exactly right, in order to obtain a good-quality image. In a system with a wide dynamic range, there is much more tolerance to camera settings. As a general rule the dynamic ranges of AgX films offer quite a wide dynamic range, and negative films have much more than slide films.

It was stated earlier that the light-sensitive elements in color negative film are the sliver halide crystals. As it turns out, all other things being equal, the sensitivity of these crystals is related to their size. The bigger the crystal, the more likely it is to be triggered by a low level of exposure – it has higher sensitivity. Smaller crystals have less sensitivity. If a film were made where all of the crystals were the same size, there would be very little response at low levels of exposure. Then as one tested increasing exposure levels, a point would be reached at which the crystals would be activated. And since they are all the same size, they all have the same level of sensitivity, and, therefore, they would all be triggered at about the same level of exposure. With increasing exposure, nothing else would happen because all of the crystals have already been triggered (see Figure 1.8).

Figure 1.8

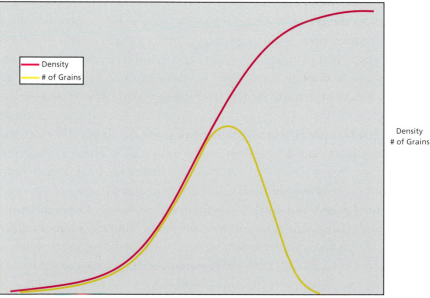

Log Exposure/Grain Size

Figure 1.9

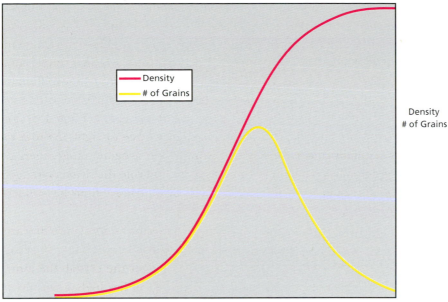

Log Exposure/Grain Size

Conversely, if there were a wide range of crystal sizes, the larger ones would trigger at lower levels of exposure, the more moderately sized crystals would trigger at higher levels of exposure, and the very fine crystals would not respond until very high levels of exposure were reached. The result is that in the first instance, there is no response at low levels, an abrupt response at some point, and no continued response after that; whereas, in the second instance, there is increasing response over a wide range of exposure levels (see Figure 1.9). Since there is a limit to the range of crystal sizes that can be obtained in any given layer, manufacturers apply two or even three layers with graduated ranges of crystal sizes to achieve a product with a very wide dynamic range and exposure latitude. This can be done in both negative films and to some extent in slide films. But, with negative films there is greatly increased opportunity to build in dynamic range.

In a photographic system, the rate of increase of image darkness in response to increased exposure is called gamma (see Figure 1.10). Gamma is related to contrast such that high gamma films have high contrast. Also, it is generally true that the higher the gamma, the smaller the latitude and dynamic range. The gamma for a final print is obtained by multiplying the gammas for the negative film and the print paper. The print will be made in the laboratory under tightly controllable conditions, so that it is not important that print papers have large exposure latitude – that is, gamma can be quite high. And, since the gamma in the final print image is the product of the gamma for the negative film and the gamma of the print paper, the negative can have a correspondingly low gamma.

Figure 1.10

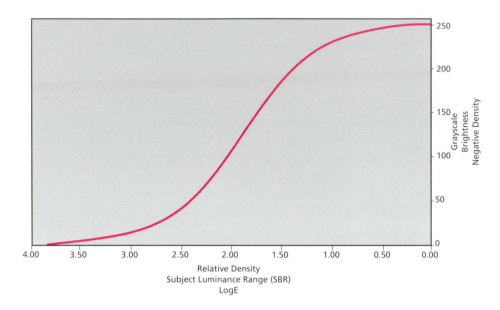

Sensitometry Response Curve

Grayscale
Brightness
Negative Density

Relative Density
Subject Luminance Range (SBR)
LogE

This can be translated into wide dynamic range and generous exposure latitude for the film that is used in the camera, where external circumstances will dictate range of light levels in the scene. This is then offset by the high gamma in the print paper, where low gamma is not important, to achieve a very robust system.

Most negative films are designed to exploit this effect as much as possible to give good results over a wide range of conditions. Note that since the negative-to-print system allows for some adjustment during the printing process, it is possible to display objects in a single print that represented a very large range of original scene illumination levels. Slides cannot do as well, since there is no printing stage. So more care must be exercised in taking pictures with slide films. In situations where the photographer is unable to reduce the exposure range in the original subject (for example in broad landscapes and crime scenes), negative films will produce superior results.

With any light-sensitive system, there will be low illumination levels, below which there is no response. That is, the response curve of output per unit of input is flat and gamma in this region is equal to zero. Then there is the area of incremental response where there is ever more output per unit input. This is where gamma has some practical value (it is a negative number for negative films and a positive for slide films). Eventually at very high levels of light, the system will saturate and there is no longer any increased output for increased input. The response curve is flat again and gamma is equal to zero again. This yields the so-called "S-shaped curve." It is possible to utilize any portion of the curve where there is some degree of slope, but all information that fell on the flat portions of the curve is lost for ever.

In summary, color negative film systems allow for very high resolution and are useful in scenes that contain a wide range of illumination levels. It is relatively easy to get high-quality prints from this approach due to masking of unwanted absorptions of dyes, and when using negative films, adjustments during the printing process. The approach is quite flexible, since different films can be used in the same camera as needed for the assignment – very sensitive films for night work, and much lower sensitivity films for work in bright settings. These products capture and display images in full compatibility with the tristimulus model of color vision, and have evolved to the point where they can accurately reproduce a very large proportion of the colors that the eye can see. These systems can operate well over a wide range of exposure times, from less than a thousandth of a second on up to hours of exposure.

DIGITAL TECHNOLOGY

Digital imaging technology is based upon the concept of dividing an image up into a very large number of very small picture elements (called pixels), arrayed in some regular pattern. This is very much like the concept followed in mosaics. Each pixel can represent only one tri-stimulus color. The smaller the pixels are compared to the full image, the finer the detail that can be shown, that is to say the higher the resolution. Each pixel must be represented by a set of discrete numbers. These tell its location in the array and the color of the light over that small area at that particular point. In film, the location is fixed by where the points are on the support sheet, and the colors are represented by the amount of dye in each of the layers at those points – there are no numbers involved.

The big advantage to digital imaging is that since numbers, or digits represent the pieces of information needed to construct the image, it is easy to use digital computers to work with the images. One can perform arithmetic functions on the numbers of a digital image, and thereby change the picture. A film image usually cannot be changed once the film has been processed, and any adjustments that one wants to make in the image must be done when the film or paper, or both, are being processed, or when the print is made. Changes require repeating the printing process, which is quite time consuming. Also it is somewhat an art to change parts of an image and not the whole image in any single process. It is impossible to make an exact duplicate of a film image because the optical steps involved always have some degree of information loss. However, a stored digital image (as opposed to a print or other presentation of that image) is nothing but a sequence of numbers, and one can duplicate that sequence with no losses. With modern high-speed computers, one can modify a digital image in mere seconds, and after viewing the result, change it back. One simple example is to change the contrast. This simply involves multiplying the color values for all of the pixels by a constant. This takes seconds to do in a modern computer, and if a readjustment is needed, it too can be done very quickly. With each passing day, more and more mathematical routines are developed to modify images and to extract information from them.

Digital images can be created by scanning film images, or other hard-copy images. This process converts a measurement of the color at each location into a set of numbers that indicate those colors at the various locations. It is also possible to use a digital camera. These devices have sensor chips located where the film would normally be in a traditional camera. The sensor chip is designed to measure the amount and color of the light at each location and convert that information into numbers that represent those values. Video cameras come in two varieties, analog and digital. The analog cameras record a signal that is proportional to the level and color of light at each point. The camera scans across the image plane in a series of rows, so the locations of the image points are deduced by knowing which row is being scanned and how far across the spot is for each instant. The result is a continuous stream of electrical signals, which vary over time, and the level of the signal at each point is an indication of the brightness of the image. Digital video cameras convert the spot color and location information to numbers, and their output is just a stream of numbers (with some header information so that the reading device knows how to interpret the data stream).

Once created, or "captured" from a live scene, digital images can be stored on many different types of media – the same as any other computer data file. For example, they are frequently stored on computer hard drives while someone is working with the pictures or as part of a computer network. For long-term archival storage, they are often stored on write-once-read-many (WORM) times compact disks. Most digital cameras store images on a flash card, a small, removable memory device. As long as the string of numbers is not altered, the images that will be produced on the computer screen or printer will not be affected by the storage medium. A digital image file is the string of numbers, which is independent of the medium holding the data. A film image is an areawise array of dye that is firmly fixed to the material on which it resides when created.

There are a number of file formats that have come into common use for images. One of the most common and most robust is the Tagged Image File Format (TIFF). In this format, the image file may be compressed a bit, or it may stay in its original form. In the original form, each pixel is individually retained and represented. There are compressed versions of TIFF, most of which are "lossless." This means that even though some information may be deleted before the image file is stored, the process employed is such that all of the original information will be returned when the file is reopened.

Another approach to compression is known as JPEG (Joint Photographic Experts Group). This is called a "lossy" compression routine because in the process of reducing the file size, some information is lost and is not recoverable when the image is opened. With JPEG, the user can select the degree of compression desired. But caution is needed, since the amount of loss increases as the level of compression increases. JPEG can be used to reduce file sizes by a factor of 100, but much will be lost in the process.

There are other compression methods, and they vary in the degree to which they reduce file size. The important distinction is that some are lossless and others are lossy. Depending upon the application, the user should choose wisely. Finally, some of the lossy compression algorithms are adjustable and the amount of loss and distortion will increase as the compression ratio is increased.

To render visible images from digital image files, one must either use a dynamic display such as a data projector or computer screen, or make a print. The display or printer will reproduce each pixel according to the numbers contained in the image file. The quality of the image will depend on the ability of the device to reproduce fine detail over a wide range of coloration.

In summary, digital technology divides images into a large number of small picture elements, creating a mosaic-like representation. Each tile in the array can assume only one color, so there is no information with detail finer than the tile size. Unlike film technology, it is currently not possible to place the light-sensitive elements one behind the other and, thereby, obtain a smooth, very high-resolution image. Also, it is not practical to utilize several receptors to extend the sensitivity range, resulting in a somewhat limited dynamic range. But, digital imaging, since it represents images as strings of numbers, greatly increases the user's ability to enhance images and extract information through the use of mathematical functions. Also, since the image is a string of numbers and not a physical entity, it is independent of medium. It can be captured using one medium, temporarily stored on another, and archived on still another medium. All of the records can be perfect replications of the original image. Images can be displayed on dynamic devices or printed onto paper or transparent material for use on an overhead projector. All of these issues will be discussed in greater detail in the following chapters.

CHAPTER 2

IMAGE CAPTURE DEVICES

An image is said to be captured when one has made a record that can be used to produce a replica of an original scene or object. There are three devices that are commonly used to accomplish this in the field of forensic digital imaging. These are the digital camera, the flatbed scanner, and the film scanner. Clearly, these are for general photographic situations, page-sized images or objects, and pieces of film respectively. Another device that is often used in forensic applications, and sometimes carries the word capture in its name, is the video capture board. This does not actually convert an original object to a record that is capable of producing an image, though. This is done by a video camera, and the capture board allows one to convert selected portions of the videotape, or stream of live video, into a digital image file. In this book, we will address the main three capture devices in some detail and just mention some of the characteristics of capture boards. We will not address digital video, however, as it is very similar to digital camera technology and it often outputs an analog signal instead of one that is digital.

DIGITAL CAMERAS

Digital cameras come in three categories:

1. point and shoot,
2. professional, and
3. high end.

The average price tends to go up by an order of magnitude as one goes up a category. Point-and-shoot cameras are available for a few hundred U.S. dollars, professional cameras tend to sell in the range of a few thousand U.S. dollars, and high-end units sell in the range of a few tens of thousands of U.S. dollars. The basic technology is very similar in all of them, but their image quality limitations and adaptability are quite different.

Point-and-shoot cameras tend to be highly automatic, have a nominal amount of image resolution, have fixed lenses, and do not accept a full range of accessories. They are designed for ease of use over a nominal range of circumstances – they make good vacation cameras. Professional cameras tend to have a full range of automatic features and a wide range of specialized settings for each, including manual overrides. They have a bit more resolution than the point-and-shoot cameras, accept many different lenses, and a full range of accessories. High-end cameras have the highest resolution of all, and some have multiple chip configurations to give especially high-quality images. By and large, law enforcement agencies have been using point-and-shoot cameras and professional cameras. They are used for first responder situations and by professional photographers respectively. A few of the larger agencies have been able to justify the purchase of high-end cameras, but on the whole, these are rare instances.

Digital cameras, like traditional cameras, have a lens to form an image in space (a real image, the kind that one can project and see on a screen), several mechanisms to control the amount of time that the lens is opened and how much light comes through per second during this time. There is also a mechanism to properly focus the lens and frame the object of interest within the field of view. Finally, there is a mechanism to measure the amount of light on the scene of interest. Many of the mechanisms, lenses, and control devices are the same ones that camera manufacturers put in their traditional cameras.

The big differences are that the digital cameras do not use film, they contain onboard image processors, and allow direct connection to a computer or a printer. Most digital cameras also have an on-camera image viewer. This is a small screen that allows the photographer to preview a small version of the image. Digital cameras also have a few additional controls to go along with the special digital image mechanisms. By and large, anyone who has used a traditional film camera will quickly recognize all of the similarities, and it will not take very long to master the additional features. But, digital cameras do not have the same performance that one takes for granted in traditional photography, and so understanding these limitations will help the photographer avoid difficulties. The best way to understand the limitations is to start with a description of how digital cameras work, and what it is in their make up that can limit performance. With this in mind, one is ready to deal with photographic assignments, ready to choose the right camera, and ready to read the advertisements to see what is new and what it might really mean relative to future purchases.

As in a traditional camera, the lens forms a real image at a plane near the back of the housing. In a traditional camera, the film would be held at this plane. In a digital camera, there is an imager chip at this location. While this is a very complicated device, the main point is that it has a large number of small light-sensitive elements arranged in some geometric array – usually just square sensor tiles arranged in a rectangular, row-by-column grid pattern. Because of the way in which these work, the imager chips are often called charge-coupled devices, or CCDs. The real image that is formed on the CCD's surface is nothing more than the concentration of a certain amount of light at each position in the plane. Dark areas in the original scene have very little light and bright areas have a great deal of light. Also, the color of the light is in accordance with the original. The small tiles act like small light meters. They respond by generating an amount of electrical charge that is in proportion to the amount of light. But this is an average across the full surface of each tile. It is not possible to respond to small bits of light or dark that are smaller than the tile. So the tile averages light and dark points on its small surface and generates charge accordingly. If there are many tiles across the width and height of the frame, then this is not much of a limitation, but as the number gets smaller, the camera becomes less and less able to produce images that contain fine detail. In film, it is possible to make the silver halide crystals very small, and to stack many of them one above the other, allowing for high resolution. Digital cameras cannot do this.

In addition to the CCD chip, the newer digital cameras have an onboard array processor chip. This receives the raw data from the CCD chip after the shutter is snapped, and converts the raw light meter readings into an image file in some standard image file format like JPEG or TIFF. The array processor is a crucial part of the system, since the raw light readings cannot be used directly to make a viewable image.

One can improve the resolution of a camera by having a CCD with a large number of pixels. Figure 2.1 shows a typical CCD. Low-end cameras, some of

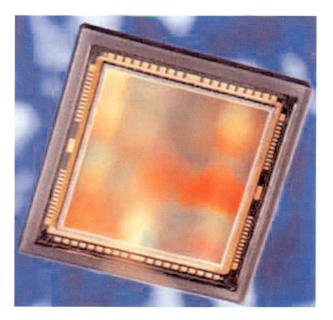

Figure 2.1

them no longer manufactured, have low ability to reproduce detail, and high-end cameras have a great deal more. In the world of film technology, engineers typically measure how many line pairs (one black and one white) a film can reproduce per millimeter. In the world of digital cameras, this is not as good a measure. With film cameras, the photographer can keep the same camera and use a sharper film type when resolution is a particular issue. This is not an option with digital cameras since the CCD is permanently mounted. So it does not matter how many lines are reproduced per millimeter on the chip. Instead what matters is what one can obtain on the image. Digital cameras range from the typical video resolution of 640 × 480 pixels per frame, on up to some 3000 × 2000 pixels. To put this to practical application, consider a photo of a latent fingerprint that has about two ridges per millimeter, and one wishes to capture the full length of the last segment of the finger, which is 3 cm long. And let us assume that to see enough ridge detail, one should be able to resolve at lease five line pairs per ridge (and trough) width. This implies that the full image should be capable of resolving 600 lines across the frame. This is obtained by noting that there were two ridges (and two troughs) per millimeter, and since it is required that there be two lines per ridge and per trough, then one needs 4 × 5 = 20 line pairs per millimeter of full image width. And since the full image of interest is 30 mm, the result is 30 × 20 = 600 lines as a minimum across the frame. One can refer to the figure to see what chip size one needs to obtain this resolution. Clearly, a video resolution camera will not give satisfactory ridge detail, although it will show some ridge patterns. Instead one needs a much more powerful camera. If the subject of interest were a palm print instead of a fingertip print, it would be hard to get a high-quality image with anything but a high-end camera.

There are two other issues that require attention. First of all, using a good-quality film one can easily get 2000 line pairs per frame width, so the limitations associated with digital cameras are not due to lenses or anything else of this sort. And while one could allude to more sophisticated measures, such as modulation transfer, the result will be pretty much the same. The data show that the predominant issue is the number of pixels on the CCD in the camera. Secondly, unlike with film, it is important to fill the frame with the subject of interest. With the film approach, there were the equivalent of some 2000 line pairs available to meet a requirement for 600. So the frame could be less than one-third filled and there would still be a good image. But with the digital camera, there is very little room for error. In rough terms, a camera with 1800 pixels across its width will meet the need for 600 line pairs, if and only if the frame were filled. For example if the frame were only half filled, there would be only 300 line pairs available for the fingerprint – half the requirement. So, with digital photography, it is important to fill the frame with the subject of interest.

There are two important technological factors that have been limiting the number of pixels on the chip in any camera. First of all, the light sensitivity of a tile is proportional to its surface area. Double the size of the area pixels, and you double the photographic speed, everything else being equal. So one factor limiting the number of pixels is that if more of them are placed on the same-sized chip, the photographic speed will go down. One might think that a simple solution for this is to increase the size of the chip. This can be done, but because of the realities of chip manufacturing processes, this greatly increases the cost of the chip. The result is that the higher end cameras have better resolution without sacrificing speed, but they cost a lot more to buy!

Another performance dimension of importance to the photographer is dynamic range. In Chapter 1, there were descriptions of the mechanisms by which film manufacturers could produce products capable of covering a rather large dynamic range. They can produce special portrait films that allow wedding photographers the ability to capture the detail in the bride's white-on-white dress in the same picture in which one can see the black velvet of the groom's tuxedo apart from the black cloth background. Digital cameras have great difficulty doing this because no method has yet been developed to use more than a single receptor tile per area on the image surface. The result is that in situations such as crime scenes or accident scenes, there are problems imposed because of the particulars of the setting. There is only so much that can be done to brighten dark areas without also brightening light areas, so the range of light levels in the scene remain a problem. As was also mentioned in Chapter 1, the eye expects to be able to see imagery over a wide range, and so deficiencies can alter the interpretation of the photograph.

Before discussing comparative performance, it is important to explain some basic principles. First of all, in the field of traditional photography, the standard for determining a film's response is to measure its density. Density is defined as the base 10 logarithm of the inverse of the film's transmission at any point. In the case of photographic paper, the papers reflectance is used instead of trans-mission. These are called transmission and reflection density, respectively. The important factor is that the density goes up as the amount of the incident light transmitted goes down, and the relationship is non-linear – specifically, loga-rithmic. This means that equal percentage changes in transmission show up as equal changes in density. So, if the reflectance is 50% (0.5), then the density is 0.3. And if the reflectance is 25% (0.5 × 0.5) then the density is 0.6. Likewise, 12.5% results in a reading of 0.9, and so on. Equal changes in the percentage change result in equal increments in density. This convention was adopted because it facilitates a wide range of readings more easily, and more impor-tantly, because it tends to better reflect how the eye sees.

In the field of digital imaging, it is more common to work with a linear measure of response called "gray level." The most common setting is, in 8-bit systems, where there are 256 gray levels – which is equal to the binary base, 2, raised to the eighth power, or 2^8. The arithmetic used in computers tends to be based upon the binary system instead of the decimal system, so the base is 2 and not 10. In situations like that of gray level, where a number is represented as a single, eight-digit, binary number (8 bits to the byte), there are only 256 combinations, and so, there are only 256 integer numbers – there is no equivalent to a decimal point. Also, if one starts with zero, then the highest number is 255, not 256. Gray levels vary linearly with the actual percentage numbers. In this case, a 100% reflectance would result in a gray level of 255, a 50% reflectance would give a reading of 127, and a zero reflectance would be assigned a reading of zero. This makes the arithmetic a bit easier. Also, since other components of a system may work with bytes with more bits, for example many cameras actually work with 10 bits, or more, it is very easy to work back and forth since the scaling is linear. For example, consider a camera working with 10 bits per number, it would register 100% reflectance as 1023 (but in binary code). The gray level for 50% reflectance would be 511, and 0% would be zero. The conversion is a simple division by 4 – the base 2 raised to the power of the difference between 8 and 10 bits (except that the arithmetic is done in binary, so there are no rounding errors as there are with the decimal representations of the binary integers).

The key point in all of this is that traditional systems are measured in logarithmic units and digital systems are measured in linear units. So, if one is to compare one to the other, a conversion must be made to one set of measurements. Since the eye tends to see more nearly like the logarithmic system, the linear data are converted to their logarithmic counterparts to make comparisons. There is one further complication, however. Transmission density is virtually unbounded on the high end, while reflection density has an upper limit in all practical implementations. Also, at the low end of the scale, reflection densities can often show an artificially high reading compared to their transmission counterparts. The upper end limitation is more important in this instance, so it will be explained.

In the case of transmission density, one is passing a beam of light through a medium that absorbs a portion of the incoming light and transmits the rest. The transmittance is the ratio of these two. In reflection density, light arrives at a surface, some of it is absorbed, and the rest reflects back to the viewer. Typically, there is a dye coated on the surface, and in reality the light strikes the surface, and most of it penetrates the dye coating. Then it hits a reflecting surface below, goes back though the dye and emerges to be reflected to the viewer. Since the light went though the dye twice, one needs less dye in the reflection mode to obtain the same effect as in the transmission mode. But the important difference is that not all of the incoming light went into the dye in the first place. Some of it reflected back right off the front surface. In all practical systems this is unavoidable and typically it is on the order of 1% or 0.5%. This means that the lowest reflectance achievable, no matter how much dye is present, is 99% or 99.5%. These may seem like low amounts of reflection, but since the eye has such a wide dynamic range, and tends to see equal percentages of change as equal amounts of density increment, the effect is very real. In practice, it is almost impossible to achieve reflection densities much higher than 2.0 or 2.3 (99% and 99.5%, respectively). So, while gray value may continue to change, it is altogether possible that the resulting reflection density does not change because of front surface reflection limitations.

In order to compare systems, measurements were made to convert gray values to equivalent reflection densities, and these produced an indication of what one would see if one were to look at a print from a digital image file. In Figure 2.2, one sees a comparison of the response curves, or sensitometric curves, of a common color reversal (slide) film and the response curve for a typical digital camera. Curves for these products are customarily drawn such that they slope upwards, but for ease of comparison, they are shown as a mirror image here. Since slides are normally viewed as transmission mode, the data have been left as transmission densities. The factor to consider in all of this is the range of light levels (measured on the logarithm of exposing light on the X axis) over which the response curve has noticeable slope. Note that where this curve is flat, there is no increase in output (density) per unit increase in input (exposure). Reading the values off the figure, it is clear that the slide film is

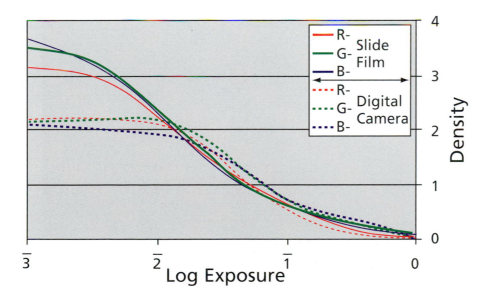

Figure 2.2
Characteristic curves of slide film and digital camera.

useful over a range of about 1000:1 (three log cycles). The negative film is useful over a range of nearly 10,000:1, and the digital camera's range is restricted to about 150:1. Even if one were to allow for using the darkest gray values by arithmetically altering the curve, these tend to be very noisy and give a highly speckled image, and the effective dynamic range is still quite small. For the ultimate comparison, the human eye can operate over a range of about one million to one.

In practice, the photographer must be aware that when using a digital camera, dynamic range will be rather limited (see Figure 2.2). So, care must be taken to use fill light for particularly dark locations, and avoid anything that might create a very bright spot, such as the reflection of a light off a metallic surface. With point-and-shoot cameras, it frequently happens that the on-camera flash will reflect too strongly off a person's face in a head shot.

One problem that has been seen in some digital cameras is color aliasing. In the introductory discussions, it was mentioned that the CCDs in digital cameras had individual color sensor tiles on their front surface. In the case of color cameras, each tile has a filter coated right on the tiles' front faces. Each tile has either a red, a green, or a blue filter between the projected image and the light-sensitive material. So, each pixel sees only one of the primary colors. But since each pixel must be represented by values for all three primary colors, provision must be made to estimate the missing data. This is done on the basis of the near

Figure 2.3

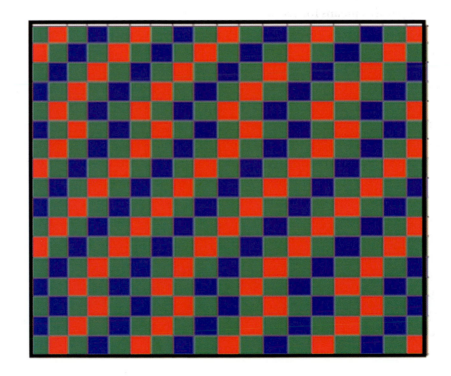

neighbors. Figure 2.3 shows the configuration most frequently used for the filter arrangement. Note that every other pixel sees green light. This is consistent with the human vision system, which gets most of the impression of image sharpness from the green record. There is less sensitivity to the other two colors. But this can cause problems if the image detail is similar in size to the pixel spacing. Technically, this is known as undersampling. There are gaps between the sensing elements – particularly for the red and blue records. For example, consider a light line between two dark areas in the image; the line is about one or two pixels wide, and it falls over a series of, say, blue pixels on the CCD. The result will be the creation of image pixels that are strongly yellow in a print. This is because the line falls in an area where the imager cannot see blue. Near blue neighbors are used to estimate the blue levels in between. But these are all in dark areas, so the image will erroneously be assigned a dark value for the blue in the area where the light line fell. This will call for a large amount of ink of the opposite subtractive primary – the opposite of blue is yellow. Hence, yellow highlight spots will be seen in the image. A similar effect will occur in the red, resulting in cyan highlights. These can be readily seen in the pictures, and they cannot be removed by color balancing, or by any other easy adjustment.

These artifacts cause serious problems. If there is a photo which includes a fine pattern on cloth, such as a striped shirt or fine-checkered jacket, its color can be greatly altered. In photos of gray hair, it is common to see patterns of yellow and cyan, distorting the true color. Recently, camera manufacturers have come up with a means for reducing this problem. A ground glass-like element is placed in front of the imager chip. Its location and degree of roughness must be carefully controlled. This device spreads the light a bit, preventing fine lines from being so fine that near neighbors are in a vastly different environment. Thus the gap between the sensors of the same color is bridged by the slightly blurred light, reducing the undersampling condition. Unfortunately, this reduces the effective resolution of the camera at the same time. To remedy this, the camera is given a sharpening filter in its onboard image processor. When combined, the two adjustments work fairly well. Aliasing is reduced, and there is a slight artificial enhancement at sharp light–dark interfaces. But the artificial edges are less objectionable than the spurious color highlights. The photographer should choose cameras carefully to be sure that aliasing is cured.

Clearly the intent in forensic applications is to reproduce images that are a fair and accurate representation of what was seen at the scene. To accomplish this it is important that colors are reproduced somewhat accurately. Any reasonable tri-stimulus system should be capable of doing this, but to do so the response curves for the three colors must have the same shape. Sliding them over the exposure axis is an easy adjustment, but bending them in differing amounts at different points is a major challenge. And, since the actual image output from the camera is computed by the array processor, one cannot assume a simplistic relationship between the original light and the responses for the CCD. If the three curves are rather parallel to each other, then it is possible to do color adjusting after the fact easily and achieve good results. If not, then most likely there will be problems. Remember that when the three curves are at the same gray level, the image will be gray at that point. If the curves are parallel to each other, then it is possible to shift them, each in toto such that they lie one on top of the other. In this situation, the image will be a grayscale – each point will appear gray. But if the curves have different shapes, then it is not possible to superimpose them (without changing their shapes) except at a single point. In this situation, the image will be gray at that point but some other color at all of the others. Figures 2.4 and 2.5 show sensitometric curves and measures of color fidelity for two different cameras. Note that the one in Figure 2.4 has a very high slope for the blue curve compared to the other colors, and correspondingly there are large errors along the blue–yellow axis on the fidelity plot. The other camera has curves that are much more alike and the error lines are very small. It will be virtually impossible to have fair and accurate representations if one uses a camera that does not have good color fidelity characteristics. The easy way to determine this is to take a photo of a grayscale under daylight conditions and see if the resulting gray value readings indicate that reproduction of gray patches in the prints will be over the full range, after color balancing.

Many of the newer digital cameras have means to do a white check. In fact, some try to do this automatically. This means that a last-minute adjustment is made by the array processor to compute a color-balanced image right in the camera. Compensation is made for the light source in this process. It does not

Figure 2.4

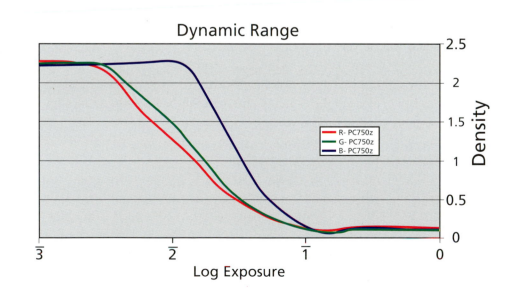

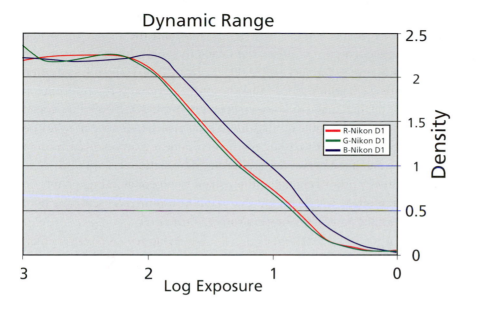

Figure 2.5

do much to help if the individual curves are not parallel. White balancing, which is common in better video cameras, is a great help, and is a very attractive feature to consider when choosing a camera.

One last issue to consider in relation to digital cameras is noise. If one were taking a photo of a perfectly uniform gray card, one might expect that all of the pixels will make the same reading and the resulting image will be a perfectly uniform gray. This is not the case. There is significant variation in the readings, one pixel to another. In many, but not the newer cameras, the effect is more pronounced in the blue channel. As a general rule, this is not a problem for the parts of an image that fall in what might be called the normal viewing light levels. And with most cameras, the effect is very pronounced for very dark positions – and will be very pronounced in long-term exposures. This means that digital cameras are not a good choice for taking photos of very dim subjects, such as alternative light source images where the intensity of the image is very low. In fact in some of the cameras, although they have settings that allow for "time" exposures of many seconds, the shutters actually close in less than 10 s. One must either get a stronger alternative light source, or use film to capture the original image.

The digital camera market is currently in a stage of rapid expansion and development. New entries appear every month or so. One camera that sets an interesting example, however, is the one manufactured by Foveon Corporation. This camera has three CCDs instead of just one – one for each primary color, which avoids the undersampling problem altogether and also avoids the need for onboard edge sharpening to compensate for a blur filter. Some of the professional grade video cameras have this arrangement, but in the Foveon camera, each of these CCDs has 2000 × 2000 pixels, providing very high resolution. Notice that with this high resolution, the image file size is huge – at least 2000 pixels × 2000 pixels × 3 colors = 12,000,000 bytes. To deal with such large images, the camera is built right on to a computer to process the information. The computer's screen serves as a viewer. Foveon claims an extended range, but images from the camera indicate that while it may be better than many other digital cameras, it is not up to what can be achieved with color negative film. But, this can be managed in a studio-like setting. This camera is intended for the professional, studio photographer, and is priced so high that it is probably not going to be acquired by very many law enforcement agencies. But, it does point the way regarding several of the current camera limitations.

FLATBED SCANNERS

Flatbed scanners, as the name implies, are used to create digital replicas of images on flat materials such as documents, photographic prints, pieces of cloth, and so on. While they do not have high resolution when materials sit above the scanning plane, they can capture images from some three-dimensional objects such as keys. They typically handle U.S. letter- or A4-sized documents. Some will accept U.S. legal-sized documents. The standard unit scans by reflected light off the front surface of an item, but it is possible to get special cover devices that will allow scanning by transmitted light. Some have the ability to hold photographic films ranging between 35 mm and 8 × 10 formats. They also vary in terms of resolution. Some will record 600 pixels per inch, while some will resolve 1200. Most manufacturers include sharpening filters with their units so that they can be read at resolutions higher than their optical rating. For example, the device might be specified as "1200 pixels per inch, 600 optical." The best recommendation is to compare units at their optical resolutions. Some record color by making three separate passes, while most of the more current units will record all three records on a single pass. One final image-related issue is dynamic range. In the reflection mode, as was discussed in Chapter 1, all units are restricted to reflection densities of 2.0 or 2.3. However, in transmission mode the units will vary one to another. They generally are in the range of 3.0 to 3.9. More is certainly better when scanning reversal film images.

Over the past few years there has been a progression of protocols under which they attach to the host computer. Parallel port connections are very inexpensive, but also very slow. SCSI connections are much faster, but like all SCSI devices, prone to problems. More recently, most manufacturers have adopted the USB interface, which is more reliable than SCSI and about as fast. For the purposes of the discussion here, the focus will be on a typical configuration of single pass design.

On the top of all units is a cover, which, when opened reveals a glass platen. In reflection mode the original is placed on the platen much the way a document is placed on an office copier platen. In some units the cover can be replaced by a document feeder, much like that on a copier, or it can be replaced by a light box, which enables scanning in transmission mode. Below the glass, there is a carriage that moves the length of the platen to make the scans. In the digital camera, there is a CCD with a rectangular array of light-sensitive pixels. Each pixel sees only one color and estimates of the unseen colors must be estimated to complete the picture. In flatbed scanners, there are three linear arrays, and usually an optical device to send the three primary color images to one of these. In this way, no estimation of missing color information is needed. And since the image is captured in a single pass, the level of precision on the drive of the carriage is quite nominal, considering the high resolution. As a result of the use of linear arrays and nominal drive precision, the cost of these devices is remarkably low. And, they are very useful devices for a number of applications.

Experience has shown that despite the indicated high resolution, a nominal amount of image sharpening after scanning is usually required. Also, one should be prepared to deal with very large image files. For example, scanning a 4 × 6 inch color print at 600 pixels per inch results in an image that is 2400 × 3600 pixels, and will produce a 25.92 megabyte file. And, a U.S. letter-sized document scanned in color at the same resolution will yield a whopping 100.98 megabyte file.

The software that accompanies the better scanners allows the operator to adjust the color, exposure, and contrast prior to scanning. This capability is often worth the cost of the better scanner all by itself. One can make similar adjustments after scanning, but a price is paid. Once the scan is complete, all of the adjustments result in a loss of some image information. The bigger the adjustments, the bigger the loss. It is much better to set the scanner for each image, and record the right information instead of trying to estimate this from extraneous information gathered in a suboptimal scan.

FILM SCANNERS

Film scanners are essentially small versions of flatbed scanners, except that they work with transmission originals and usually at higher resolution. Typically these devices operate between 2000 and 2800 pixels per inch. Many of them are three-pass scanners, recording each of the primary color records on each pass. Most accept negatives and slides, and some have film strip feeders. All accept

35 mm film, and some will also accept larger formats. Note that the file size at 2800 pixels per inch will be 2800 × 3500 pixels, essentially the same as scanning a 4 × 6 inch print made from the same negative. File sizes are large, on the order of 20 megabytes for a 35 mm frame, and much more for larger formats.

It is instructive to compare the scanned negative to the digital camera. The high-end cameras usually are limited to a 2000 × 3000 pixel image. Also, since the film is able to record information over several orders of magnitude of initial exposure, and it is possible to retrieve all of this information in the scan, the scanned negative will have significantly more information than even the better digital cameras. Also, film has excellent color reproduction, there is no color aliasing issue, and there is very little dark area noise. The overwhelming choice when high-quality images are required is to use film in the camera, followed by scanning if it is important to extract measured information from the photos. Digital cameras gain their advantage in situations where instant viewing of the image is highly desirable, and second-tier quality is acceptable, and/or there are ways to circumvent the limitations.

Film scanners usually come with software that allows adjustment for the particular type of film being scanned. These are listed by brand, type, and ISO. There are also more generic settings to use in case the details are not known. Typically, a bit of edge enhancement after scanning is a good idea. As with flatbed scanners, a bit of attention to optimizing the scanning parameters prior to the actual scan will produce much better results than if one tries to correct the image afterwards.

One question that frequently comes up is, "Should I scan the negative, or scan the print?" Most of the time the answer is, scan the negative. As was seen, the pixel counts are comparable, so pixel-limited resolution is about the same. But in making a print from a negative, some of the information is lost due to imperfections in the optics. Also, there is the issue of front surface reflection limitations. The maximum range that the print scanner will be able to record will be limited to density readings of 2.0, or possibly 2.3. If the original scene had information that was greater than this 100:1, or 200:1 range allows, that information will be lost. But, the negative is scanned in the transmission mode, and most film scanners are able to record over a range of over 3.0 density – that is, a few thousand to one. And the film was able to capture over a range of several thousand to one. So, scanning the negative will allow the capture of a great deal more information. If one has a 5 × 7 inch print instead of a 4 × 6, and if there is not much information in both shadows and highlights that must be captured, or if one scans at pixel resolutions in excess of 600 per inch, it might be just as effective to scan the print.

Earlier it was mentioned that one could scan prints or films at differing resolutions. And, when higher resolutions are used, the files can become very large.

Accordingly, it is considered good practice to "scan for the printing device." This means that when one should anticipate how the image would be printed and then set the resolution to match that expectation. If one is going to print to a film printer (a device that writes digital image files to film), and if that will be done at 2500 pixels per inch, then the scan should be able to match that resolution. Whereas, if the image is going to an inkjet printer that is operating at 300 pixels per inch (not to be confused with dots per inch, which gives a much higher number and is how many printers are specified), and there will not be any enlarging of the image, then one only needs to scan accordingly. On a flatbed scanner, this will probably mean either 300 or 600 pixels per inch. Film scanners are less flexible in this regard.

Experience has shown that a person, looking at a paper print, held at normal viewing distance will be quite satisfied if the print has 200 pixels per inch. If the image is 6 inches across, then the image will require $200 \times 6 = 1200$ pixels across its width. If the same image is to be printed at 4 inches, then only 800 pixels are required. In some cases, it is expected that the viewer might look at the image more closely, studying detail in a particular portion of the image. Under these conditions, 300 pixels per inch will be required. If closer detail is required, then the image should be reprinted from an enlarged file. So, to set the parameters for scanning, consider the printing. To make a print of a certain number of inches across, multiply that number by either 200 or 300 (depending upon expectations) and scan to achieve the proper pixel count. Often, when scanning negatives, one plans to archive the image on a CD, in which case it is not altogether clear what viewing conditions may be sought in the future. In this situation, the rule of thumb is to scan a 35 mm frame for about 20 megabytes. Larger formats should be scanned to proportionately larger file sizes. This will allow for some enlargement later on.

CHAPTER 3

IMAGE OUTPUT DEVICES

When working with digital imaging, one has two main options for viewing images: dynamically displayed by the computer or as hardcopy prints. Dynamic display of a direct view screen is generally used to provide feedback to the operator regarding actions being taken, and dynamic display via a data projector onto a large viewing screen is usually the choice for sharing information with a larger audience. As for printing, there are several options, but usually the objective is to have the ability to view images without any active equipment or to facilitate transportation. Sometimes a combination might be used. For example, some trial lawyers will show images to a jury as a whole using a data projector and a large screen. At the same time, individual, page-sized prints are provided to each jury member for more personalized viewing. There are a number of factors associated with each of these options that should be of interest to the forensic user. In general, however, they break down into issues of resolution, dynamic range, and color gamut – the range of colors that can be produced.

DYNAMIC DISPLAYS

There are many technologies now being used as dynamic displays. Some of them work with predefined pixels and some do not. Also, some have screens that are viewed directly while others use a projection approach. A few of these will be reviewed here in order to highlight some of the key performance issues. Table 3.1 shows how the technologies sort among the categories just mentioned.

DIRECT VIEW TECHNOLOGY

These are typically the primary devices used by a computer to provide feedback and to communicate with the user. They are generally intended for a single user sitting within 18–24 inches from the viewing screen. The plasma display units are relatively flat and so take up less desk space. The CRTs are much larger and

Table 3.1

	Predefined pixels	**No predefined pixels**
Direct view	Plasma display	Cathode ray tube (CRT)
Projection	Liquid crystal display (LCD) Light processing	Light valve technology (LVT)

heavier. CRTs can provide more brightness than plasma displays, as a general rule, and so might be better in areas of high ambient light. It is important to note that the color of the screen when the unit is turned off is "black." The device adds brightness in the brighter areas, but it cannot subtract brightness. So the darkest areas, that is the black portions of the image being viewed are limited to the color of the screen when its power is off. In situations where there is high ambient light, the dark portions are unacceptably light. Or if there are glaring light sources reflecting in the screen the view is severely distorted. A few manufacturers provide screens, similar to window screens. These are coated with a dull black paint so as to minimize reflections on their front surfaces, and they also provide some shadowing of high ambient light. But, since they are mostly of open area, they do not block much of the light from the display device. The result is that they help to produce satisfactory "blacks" even in the presence of high ambient light. The amount of brightness a screen can generate is related to the amount of current that can be sent through each element of the device screen, and generally the newer devices can achieve satisfactorily high levels of light.

There are four key parts to the device. First of all, there is a set of three electron guns, each of which emits a stream of electrons from a structure at the rear of the tube. The intensity of the stream is under constant control – an increase in the number of electrons per second (that is, the current) results in an increase in the brightness seen on the face of the screen. Secondly, there is a deflection system. This is controlled by signals from the computer and directs the stream of electrons to a particular location on the screen. In general, this is a sweep pattern that goes across the top of the screen first, then "flies back" to the original side and down one row, only to scan across again. This process repeats on down the face of the screen to complete a single frame. When the frame is completed, the spot is directed back up to the upper row again, and the process repeats once again. On the inside of the face of the screen there is a pattern of phosphors. These produce red, green, or blue light when struck by electrons (CRT displays work with the additive primary colors). The more electrons per second, the brighter the glow. They are arranged in small clusters such that the viewer sees each cluster as a single entity. Finally, there is the shadow mask. This is in front of the phosphors and has small holes etched through it. Each acts like a small window and assures that the right amount of each phosphor shows through in the right location. The front of the mask is black and helps the screen appear dark when the phosphors are not emitting light. There are two common arrangements for shadow mask patterns. One is a series of vertical lines and the other is a set of three circles. Hold a magnifying glass to your screen and look at a bright gray or white spot. You will see the glowing phosphors and the pattern of your screen's shadow mask. Figure 3.1 shows the typical gamut for a CRT computer screen. The points of the triangle are the points at which each of the three phosphors are located when illuminated individually. All of the points inside the triangle are the result of combinations of the three. Note that there are many colors that your eye can see that the screen cannot produce. In PhotoShop software, you will see a warning if you try to work with a color that is "out of gamut."

Screens are rated in terms of their pitch, which is the distance between clusters of three spots in metric units. So, for example, a screen might be specified at 0.027 cm between cluster centers. This is roughly equivalent to 92 clusters/inch of screen, which means that the screen is capable of rendering images that have up to 92 pixels per inch. Another common standard screen resolution is 72 pixels per inch. The smaller pitch numbers mean that more detail can be represented on the screen. In addition, with most monitors, the user can adjust the software to control the equivalent number of pixels that will be presented on the screen. The most common configuration is 800×600 pixels, but there are others that are used in certain applications. Some of the

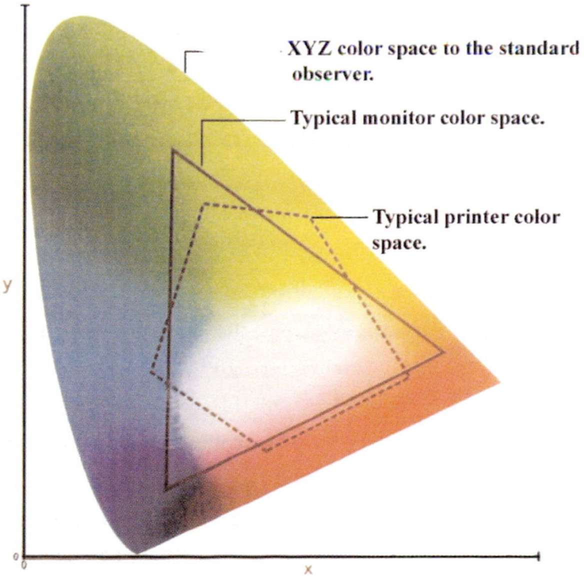

Figure 3.1

XYZ color space to the standard observer.

Typical monitor color space.

Typical printer color space.

common settings, shown in comparison to digital camera resolutions are shown in Table 3.2.

Table 3.2

Screen resolution	Pixels	Digital camera resolution	Pixels
1024 × 768	786,432	3000 × 2000	6,000,000
848 × 480	407,040	2225 × 1483	3,300,000
800 × 600	480,000	1775 × 1183	2,100,000
720 × 480	345,600	1500 × 1000	1,500,000
640 × 480	307,200	1225 × 816	1,000,000
		640 × 480	307,200

At the same time that the higher resolution setting will show finer detail on the screen, it will make the various icons and text items appear smaller. These are represented in their respective files by a fixed number of pixels. If there are more pixels per inch on the screen, then the item, with its fixed number of pixels, appears as a smaller portion of that screen.

While the preceding paragraph spoke in terms of pixels, the reference is really to equivalent pixels. CRT displays do not have fixed pixels. Instead the beam of electrons from the gun is both swept across the screen face and modulated in terms of the number of electrons per second being sent. The file being displayed is stored as a set of pixel values, and this information is used to modulate the electron beam as a function of time. Simultaneously, the sweep generator driving the beam across the screen face has its velocity adjusted. The result is that the number of pixels in the image is a function of what is in the file, and the number of equivalent pixels seen on the screen is a function of the beam modulation and sweep rates. So, when one changes the settings from 800 × 600 to 1024 × 848, one is really changing these two settings. Many older screens are not capable of the higher settings.

For forensic work it is strongly recommended that one work at higher resolution rather than lower. When studying large numbers of images on a screen, it is much more comfortable when it is easy to see detail. For purposes of comparison, it is common in the graphics industry to consider a printed image with 200 pixels per inch to be satisfactory for normal viewing. If one might study the image by holding it closer than normal viewing distance (18–24 inches), 300 pixels per inch is recommended. For most images, one can go as low as 100 pixels per inch and not be particularly aware of the lower resolution except for discrete items such as text. Notice that the 92 pixels per inch limit on even the best screens is at just about this point. Which is to say that even the best screen images would be borderline if they were prints. When the 72 pixel per inch settings are used, there is noticeable loss of quality, and one will do a lot of onscreen enlarging, which can get quite tiresome over long periods of time.

It should be noted that computer monitors use the same basic technology as television screens, but they are not the same devices. Computer monitors have significantly higher image quality. One can purchase a device to connect a TV to a computer, but if there is fine detail, or text in the image being shown, the result will probably not be very satisfactory.

PLASMA DISPLAYS

Plasma displays are technically quite different from CRTs, but much of the performance discussion above will apply. These devices comprise a large number of very small cup-like structures. Each has an emitter through which electrons are released into the cup. The surfaces are coated with a phosphor and when the electrons strike the coated surfaces, they glow. Cups are arranged in sets of three – one for each of the additive primary colors, and unlike CRTs each set of cups constitutes a pixel. Since there is no swept beam that can build equivalent pixels as it moves across the screen face, these devices are considered to be formed of discrete pixels. The computer can still apportion the pixels in the original image file to the set on the screen, but the screen's pixels are fixed in place.

In addition to being much thinner than CRTs, plasma displays lend themselves to having a flat screen instead of the convex surface one finds on most CRTs. CRTs, with their electron gun located centrally at the rear of the box, are much easier to build with a convex surface. If the surface were flat, then the electron beam as it sweeps across and up and down the screen would have to travel further at the edges of the screen than at its center, and the angle at which the beam strikes the phosphors would change as well. These factors would cause the image to be significantly brighter in the center than at the periphery. It is possible to make corrections for these effects, but it adds to the cost of the device. Plasma displays, on the other hand, comprising an array of small cups, do not have to accommodate beam-sweeping effects, and so can easily be made flat.

DATA PROJECTORS

Another way to see images while they are active in a computer to is use a data projector. These devices are used in a fashion similar to a conventional slide projector, but there are no film-based slides. The projector is connected directly to a computer, and what would normally be seen on the computer's screen is projected onto a large screen for several people to view at the same time. This technology is now finding applications in briefings, training, and in courtroom presentations. As with direct view displays, these devices can be sorted into those with discrete pixels and those without.

LIGHT VALVE PROJECTORS

These are basically analog devices that can be driven by digitized signals. Imaging is carried out by three small, monochrome cathode ray tubes. These are used to create the basic image, one each for red, green, and blue. The surface of each tube, however, is coated with a layer of liquid crystal material. This material responds to the sweeping electron beam by changing its reflective properties. The material is a circular polarizer – light that is polarized at the correct angle coming in is reflected off, and light at other angles is absorbed according to how far off the angle of polarization. The result is that the exiting beam is polarized. If this beam is then passed through another, transmissive polarizing filter, then the final beam will be darkest where the angle of polarization was farthest from the correct angle. The end result is that the sweeping electron beam has the ability to modulate a separate, broad beam of light. To complete the package, bright light sources are employed, and each is housed so as to produce a broad beam of light. This passes through a color filter and then impinges on the CRT/liquid crystal surface. The reflected beam has some elements that are polarized at the correct angle and others that are not. This is in accordance with the sweeping electron beam. Finally, the reflected beam is sent through another polarizing filter and a lens is used to form an image of the CRT face onto a projection screen. The same is done for the three primary colors and the exiting image beams are aligned so as to produce a superimposition of the three beams. Thus the color image is produced. The use of liquid crystals and a separate polarizing filter facilitates the use of very high intensity light, and so these projectors are able to work well in large auditoria. They are large devices and often are suspended from the ceiling of the room. As with the other CRT technology displays, these devices do not have mechanically defined pixels, so pixels are formed electronically.

FIXED PIXEL DATA PROJECTORS

While there are a few different technologies currently being used in devices, two of these are expected to be the leaders. One of the technologies is based on a triple pack liquid crystal and the other is based on a large array of very small, movable mirrors.

LIQUID CRYSTAL PROJECTORS

New materials and manufacturing processes have enabled development of a thin stack of three separate liquid crystal filter layers. Each passes one of the additive primary colors and is independently controlled. Light from a bright source is passed through the stack and then through a polarization filter. Then the beam is directed through a lens and onto the screen. The device is relatively simple and can be made light and compact. Many of the portable projectors use this technology. Recent advances have made it possible to obtain rather high levels of light from somewhat compact devices. The liquid crystal material is coated in discrete sections to form the pixels.

LIGHT-PROCESSING ARRAYS

These devices depend upon a very special device – an array of small, movable mirrors. Each mirror is mounted in such a way that it can be tilted in response to a control signal. Also, each mirror is one pixel. The overall configuration of the device has a bright light which impinges on the array of mirrors. From their surfaces, the light beam continues through a rapidly rotating filter wheel, which has red, green, and blue segments. The movement of the wheel is synchronized with the control signals such that the blue values for the set of pixels is imposed at the same time that the blue filter is in the beam. Then the green values are imposed and the green section of the filter wheel comes into the beam, and so on. This process repeats very rapidly so that the viewer sees the three records as if they were all present at the same time. The device is capable of vibrant colors and very high levels of brightness. The units are relatively light and compact, and so a good choice for a portable projector.

PERFORMANCE ISSUES

Data projectors are rated, first and foremost, in terms of the amount of light they can project onto a screen. The common measurement is lumens, which is a measure of light flux and can be thought of as the total number of particles of light (photons) per second. If one were to divide this by the area of the image, the result would be a measure of intensity, or brightness. So, the more lumens, the brighter the screen will appear for a given size of image. Making the image larger, either by moving the projector further away or by changing the magnification on a zoom projection lens will make the image less bright, given the number of lumens a projector can generate. For courtroom use, it is desirable to have at least 1000 lumens, and 2000 is even better. At the higher level, the courtroom can be at nearly normal illumination and the image can be fairly large. At the lower level, it may be necessary to dim the room's lights, or reduce the size of the projected image. The units with the higher lumens tend to require a substantial fan to cool the light source and hence tend to make a bit of noise. However, a good-sized room can usually absorb that noise with no problem. It is usually possible to have the computer monitor (or laptop screen) and the projector showing the same material at the same time. This allows the user to look at the audience, glancing over the top of the computer, and speak with confidence about what the audience is seeing on the screen. This makes for very effective presentations since the speaker can "read" the audience as he or she is speaking.

Another key measure of data projector performance is called contrast ratio. To make this measurement, one would project a fairly coarse, black-and-white checkered pattern onto a screen, then go to the screen and measure the incident light level in the black areas and the white areas with a light meter – the bigger the ratio the better. High contrast ratios allow the projection of bright, vibrant colors. Ratios of at least 100 to 1 are good.

Another performance indicator to consider is the uniformity of the projected image. To test this, one might project a uniform, light gray image onto the screen, then measure the incident light level at several points on the screen, especially looking to compare levels near the center with those near the edges. The difference between the readings indicates the extent to which the projector suffers from non-uniformity. Ratios approaching or exceeding 2 to 1 are bad.

When attaching a data projector to a computer, it is important to be sure that both are set to the same screen resolution. For example, if the data projector is working at 800×600 pixels, then the computer must be set to that level as well. Most data projectors come with software that facilitates this adjustment – in fact, some actually make the adjustment automatically.

Finally, the lamps used in data projectors might be very special and very expensive – especially the high-brightness units. They also have a very limited life. Accordingly, it is strongly recommended that at important presentations, there be a relatively new bulb in the projector and a spare on hand! Since these bulbs burn at very high temperatures, and the whole lamp assembly gets hot, it may take quite a while for the unit to cool down sufficiently to change a bulb. Each projector will have a useful bulb life designation.

PRINTERS

There are five types of printers currently popular in forensic imaging use: inkjet printers, dye sublimation printers, laser/xerographic printers, photo paper printers, and film writers. These are approximately in order of prevalence of use. The technology employed varies greatly across all of these types, but there are three issues that should be made clear for all. These are:

1. the number of pixels per inch,
2. pigments versus dyes, and
3. the difference between dots and pixels.

PIXELS PER INCH

It is not hard to show that an image of more than 200 pixels per inch, viewed at normal viewing distance has just about all of the resolution that a person will be able to distinguish. If someone wanted to look more closely, the requirement might go up to 300 pixels per inch. Images that have less than 100 pixels per inch suffer from visibly distinct pixels (pixelization), especially in sharp, high-contrast transition areas that are not vertical or horizontal. An example would be text. To expand this, consider an image projected onto a screen. In the paper example, one is dealing with 200 pixels per inch over an image that is about 10 inches across and is being viewed at a distance of about 20 inches. This image has 2000 pixels across its width. Extending this to the projection setting, the viewing distance is more like 15 feet, or 180 inches, and the image expands to 90 inches across. So, the pixels per inch required to give the same quality of image is now 22.2. Note that this will give the same number of pixels across the full image, $22.2 \times 90 = 2000$ (allowing for rounding off) – the same as the paper-based image. Extending the situation in the other direction, where the person looking at the image might hold it up closer to look for some small detail, another factor comes into play. The viewer is not really looking at the entire image – only a portion of it. Say the person decreased the viewing distance to 13.3 inches. (Most people do not feel comfortable looking at things that are closer than 10 inches.) In this case the image will need some 300 pixels per inch in order to give an equally satisfying picture. This exercise is intended to illustrate the rule of thumb that a printed image should have between 200 and 300 pixels per inch, and also to show how one can extrapolate this guideline to other viewing conditions.

Another factor to consider is the size of the image relative to its pixel resolution. It has just been shown that 200 pixels per inch is a good resolution for a printed image. Also, 300 pixels per inch is a reasonable upper end for this parameter. In fact, very few printers, including some of the better ones, cannot print more than 300 pixels per inch. What if one were to scan an 8×10 inch photographic print at 600 pixels per inch. The resulting image file will hold all of this information. If one were to send the image to a printer, and retain the original image size, half of the pixels will not be needed. In fact the printer, or printer driver software will probably combine adjacent pixels to create a revised image that has only 300 pixels per inch, and then print that. Alternatively, it could try to print it as a 16 × 20 inch image instead of the original 8 × 10. But some adjustment will have to be made. The reverse could occur as well. Suppose that the original is a driver's license photo that is 1 × 1.25 inches. And, suppose that the operator wants to make 4 × 5 inch prints to show to possible witnesses. If the image is scanned at the same 600 pixels per inch, and printed at 300 pixels per inch, then the resulting image will be only 2 × 2.5 inches. In order to get the size up to 4 × 5, and keep the 300 pixels per inch resolution, new pixels will have to be created by estimating their values from adjacent pixels in the original image. This process of rescaling the image is called interpolation, and it is said that one can interpolate an image up, that is add pixels, or interpolate it down by condensing pixels. The process of estimating the values for the new pixels is called resampling. There are various mathematical functions for doing this, but the most common is called "bicubic." This gives satisfactory results over a wide range of applications. A rule of thumb that is commonly followed is that one can interpolate up by a factor of two without serious problems, but going much beyond this – up by a factor of two again, creates more estimated pixels than true ones. The image is becoming falsified.

DYES VERSUS PIGMENTS

In normal use of the terminology, dyes are colorants that are effectively dissolved in whatever medium holds them. They absorb certain colors of light, but they do not deflect light rays in the process. The typical phrase is that they are transparent. If one were to look through an optically flat coating of a dye the image would be the same as if it were not there, except that certain colors would be brighter. By comparison, a pigment is a dye that has been added to a distinct particle. The particles might then be coated onto a surface and held in place by some transparent medium. But pigment particles usually have distinct, and often rough surfaces, so some of the light passing though the coating will be deflected. So pigments not only absorb certain colors, but they tend to deflect some light as well. In paints, where a high degree of coverage, that is, opacity, is the objective, pigments are an excellent solution. They scatter much of the incoming light and reflect it back to the observer. The underlying surface is covered and cannot be seen. Paint colors are obtained by blending pigment particles of several different individual colors together to get the appropriate final color.

But, pigments are not the best choice for printers. In a printer, one wants to be able to individually add dyes representing the three primary colors and in this way, create specific colors point by point over a large surface. It is easy to do this in layers of the subtractive primaries, cyan, magenta and yellow. But if a layered approach is used, it must be possible to see through each of the layers to those beneath. This can easily be done with dyes, but it is hard to accomplish with pigments. In particular, if one wants to use the print as a transparency by coating it on a clear sheet instead of paper, the deflection of light associated with pigments will be a problem, resulting in fuzzy images on the screen.

To print with pigments, one must create small dots and place these close together so that they appear to be a single unit, or pixel. In color systems, there will be three sets of dots in the small location. It is easier to envision the process in black and white, however. If we want a printer that is capable of 64 levels of gray, going from white to black, then an array of 8 × 8 dots might be a good configuration for a pixel. If none of the dot areas is filled with pigment, then the pixel is white. If one dot area is filled, the pixel is slightly darker. And as more and more of the dot areas are filled, the pixel gets darker and darker until they are all filled and the pixel appears black. In this example, it would take 8 dots per inch to yield 64 gray levels and a resolution of one pixel per inch. Similarly, a printer with 800 dots per inch would be required to achieve 64 gray levels and 100 pixels per inch. If more gray levels are desired, it will be necessary to have even more dots per inch. So, if one were seeking 121 gray levels, the printer would have to produce 1100 dots per inch to achieve a print resolution of 100 pixels per inch. To achieve 8 bits, or 256 gray levels would require 1600 dots per inch!

So, comparing printers on the basis of specifications is quite complex, and frankly most manufacturers do not give you all of the data you require. So, if the printer is a dye-based printer – and only the dye sublimation printers in the listing above are dye based – the layers superimposed, and the darkness of each spot is controlled by how much dye is deposited as opposed to where it is located. Hence dots and pixels are the same thing. A listing of 300 *dots* per inch is also 300 *pixels* per inch. For all of the others, the ratio is not one to one. And since the manufacturers hardly ever indicate how many gray levels they can produce, the exact relationship is not decipherable from the data provided. In pigment-based printers, the number of dots per inch will be very high compared with the lower limit 100 pixels per inch in order to achieve a significant number of gray levels. In pigment-based *color* printers, dot patterns for cyan, magenta, and yellow are partially superimposed, and the patterns are at different angles to each other. Also, most pigment-based color printers use undercolor replacement. So when equal amounts of cyan, magenta, and yellow, are called for, they can apply an appropriate amount of black dots instead. This not only saves colorant, but it simplifies the dot patterns and produces more satisfying prints. This approach is referred to as CMYK. The newer inkjet printers make use of more than three primary colors. Just as the use of black to replace combinations of the primary colors can simplify the dot pattern, other blends of two primaries can help as well. The result is that even with less than 256 levels for each color, one can achieve the equivalent by using more colors. This results in higher pixel resolution per dot size and gray level requirement. The newer inkjet printers, printing onto high-gloss, absorbing surfaces, can achieve excellent prints. Xerographic printers cannot take advantage of this approach as well, and so suffer from lower image quality.

Photo paper and film printers sweep laser beams over the surface of either a piece of photographic paper or film and modulate the beam as it moves. The material is then chemically processed as normal. Paper printers are capable of very large prints with very high resolution, and film printers are used to make slides for presentation in a slide projector, or in a different set-up to write to motion picture film to make animation effects movies. These are both capable of very high-quality imagery. In law enforcement work, paper printers have been used to make large courtroom displays that can also be viewed somewhat close up. An example would be large prints of aerial photos of a crime scene and venue. These machines are very large and very expensive, however, and so their use is not widespread.

CHAPTER 4

STORING AND ARCHIVING IMAGES

Compared to word processing and spreadsheet documents, images generally result in very large files that consume large amounts of file storage space, and they are much more vulnerable to manipulation than silver halide (AgX) images. Hence there is a need to develop and follow data storage procedures that will minimize these potential problems. The key issues to consider are:

- Stages of images
- Choice of storage medium
- File formats
- Current file management
- Legacy file management

STAGES OF IMAGES

In forensic digital imaging, it is useful to consider six different stages of images. These are:

1. Primary image – the logical record that results from initial capture.
2. Original image – the first fixing of the primary image to a physical medium.
3. Archive image – an image that will be kept for an extended period and recalled as needed in the future.
4. Working image – an image that is the subject of image processing.
5. Final image – the data record that is used to produce a visible image that is used for forensic purposes.
6. Presentation image – a visual manifestation of a final image (either in hardcopy form or on a display screen).

In this chapter, these designations will be described and the practical implications for each will be clarified. It is helpful to start with the traditional silver halide image and recall some of the above issues in this context. Many of the concepts are not new, but the realities of digital imaging are such that the demarcations need to be considered so as to assure proper use in various situations.

In traditional AgX photography, a latent image is formed on a piece of film. The image is not yet visible, and it can easily be modified using double exposure techniques. In addition, it is possible to adjust the processing in certain ways to achieve different effects in the image that is ultimately developed. For example, if one were to greatly reduce the level of agitation in the developer bath, the result would be the creation of fine, bright lines around dark objects. Also, temperature and processing time adjustments can change the contrast, the darkness of the image, and the graininess. The result is that there are ways in which silver halide latent images are vulnerable to alteration as are digital images. Once it is processed, however, the film image is no longer as vulnerable and one can hold it in one's hand. The image and the medium on which it resides are the same physical entity. As such, the latent image has some of the properties of a primary image, but it is normally fixed to a physical medium, and thus could be called an original image.

Early stage digital images are different. The image is really just a data file. It will happen to reside on some sort of medium or other, but the file can be duplicated, or replicated, making an exact second-generation image file that is indistinguishable from the first generation in terms of the ability to produce a viewable image. And in most cases, the first-generation file will be erased before long and the medium reused. The only difference between the generations is the medium holding the data. The image itself is not a physical thing – it is a logical thing. It is a very highly structured string of data. While an image is in this form, it is very ephemeral and vulnerable. In many ways the first-generation image is analogous to the latent image, but it is different in at least one key way – the lack of being fixed to a medium. To deal with the matter, the first-generation image in digital imaging is called a "primary image." It is used to hold the result of the data capture process, but will not be kept in this form very long. It, like the latent image, needs to be fixed to a particular piece of physical medium. The first fixing of an image results in the creation of an "original image."

The best way to create an original image is to duplicate the primary image using a blind replication tool. For example, if the primary image file is duplicated as an unopened file, and *fixed* onto another medium, one can create an original image. In this way it can safely be said that the original image had not been edited in any way prior to its creation. Note that the primary image and the original image are in the same file format. It is a faithful duplicate of the primary image in every respect except for the medium on which it resides. The result is the conceptual equivalent of a silver halide negative.

Once original images are formed, working copies can be generated. These are initially replicas of the original and primary images (both), but these can be opened and edited. As they are enhanced, they cease to be duplicates, and if it is expected that there may be some challenge to the final images, care should be taken to record all of the enhancement steps. In this way, a duplicate of the original image and the series of steps can be given to an independent person trained in the field, and they will come up with the same final image. This will go a long way to relieving any accusations of unfair manipulation.

There are some techniques, such as watermarking, hashing, and the use of embedded codes, but these, in their current state can be more of a problem than they are worth. They are defeated as soon as an image is processed, and they often are not applicable to devices such as film and flatbed scanners. And, different manufacturers use different, proprietary techniques. Simply fixing a primary image in a controlled fashion to create an original defeats most of the problems.

Another issue that was considered when the "primary image"/"original image" nomenclature was developed has to do with the perception of the ordinary meanings of the words and the realities of digital imaging technology. Digital cameras use removable and reusable recording media for the large part, and the economics of the situation are that the media will indeed be erased and reused. When images are captured using a scanner, they are typically captured on a computer's hard drive – and the hard drive space will have to be erased and reused in due course. As a result if one were to call the first assembly of the image data the "original image," one would have to deal with questions put before lay people regarding why the "original image" was erased. Since it is necessary to separate the first data assembly from its replication on a fixed medium, it was decided to call the first instance the primary image, and the second, the original.

CHOICE OF MEDIUM

While there are many options to consider for archival storage of digital images, especially original images, one medium stands out today – the write once read many times (WORM) CD with serial number. One manufacturer of these disks currently has sole use of a polymer that affords much longer life – stated as 100 years.

These disks can hold over 600 megabytes of data, which means that in most instances they can hold all of the materials associated with a single case. They cannot be erased. Once something has been written to them, it cannot be removed. When writing these disks, it is possible to "close" them. It is not possible to add information to a closed disk. The piece of medium is unique – the physical disk has a unique serial-number identifier. These disks are relatively inexpensive and it is easy to write them at a personal computer.

Figure 4.1

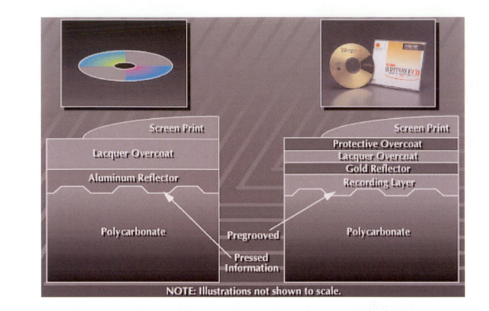

WORM CDs have four layers that are important in understanding how they work (Figure 4.1). First of all there is a plastic substrate. This is designed to provide sufficient strength to make them easy to handle and to not suffer from so much temperature-related size change that there is a danger of their not being readable. On top of this layer, there is a gold reflecting layer. Gold does not tarnish, and so the integrity of the reflecting layer will last over a very long time. Next, there is a recording or dye-polymer layer that is modified during the recording process. The dye in this layer absorbs light in the infrared portion of the spectrum, and with sufficient power input, the polymer will deform in response to the light energy. Finally, there is an overcoat on the top to protect the recording layer and assure the integrity of the information over long periods of time and many handlings.

To write to one of these disks, a high-intensity laser beam is used, with its color selected to match the absorption peak of the dye in the recording layer. The laser is driven through a switch, which in turn is driven by a counter. A clock generates pulses at regular intervals. The counter keeps track of how many have gone by since an initiation pulse. The initiation of the counter places a number in a register. When the counter has reached the given number, the switch is thrown and the counter is reinitiated with a new number. The process repeats over and again at very high speed. The switch controls the writing laser. When the switch is on, the laser is on, and conversely when the switch is off, the laser is off. When the laser is on, its beam heats the recording layer to the point where the material deforms, creating a hollow space at that location. During all of this activity, the disk is spinning at a constant rate. In this way, the on-phase results in creation of a trough in the recording layer, with a length that is proportional to the number in the counter. Immediately following the trough there will be a portion with no deformation. Again the length of this land portion is proportional to the number that was in the counter when it was active. So, the written disk contains a series of troughs, following a spiral track, with lands between the troughs. The lengths of the troughs and the lands are each proportional to a specific number. The sequence of numbers is the data in the file to be recorded.

For short-term file storage, there are many options. The best advice in choosing among these is to stay with widely used media. Some digital camera manufacturers have a removable memory device that is essentially unique to their products. A few others use a flexible memory device, which begs the question of durability. Recently, some manufacturers have used 1.44 megabyte floppy disks. This forces the use of a large degree of image compression that is likely to significantly alter the images being recorded. An image from a 3.3 megapixel camera will result in an (uncompressed) file of 9.9 megabytes. For practical reasons, one might want to hold some 10 or more images on a disk. Since each disk can only hold about 1/7 of an image, and there might be 10 images on the disk, each image must have been compressed by at least a factor of almost 70 – a significant portion of the information that is not recorded is lost forever, and unwanted artifacts will be inserted into the image.

Solid-state flash cards have sufficient capacity to store many images without excessive, lossy compression and they are rather durable in the field. This is the choice of most of the digital camera manufacturers. Some of the older digital cameras use PCMCIA cards, which are essentially small hard drives. These are a fine choice in all but two respects: they are fragile and they are expensive.

With most scanners, the primary images are recorded directly onto the host computer's hard drive. In some cases an external drive of some sort might be used, but it is very likely that the medium is a magnetic memory device of some sort. As with all primary images, these should be fixed onto a permanent medium as soon as possible. Note that with scanners, it is often not possible to copy the images in an unopened state. At the same time, scanners all record some sort of physical item that was in them when images were captured, and if that thing can be preserved, it serves to protect the integrity of the process. If it can't be preserved (for example the scanning of a questioned document that will then be subjected to a ninhydrin treatment), then the issue will have to fall back to the operator and the processes followed. In any event, it is not a good practice to leave primary images on the internal hard drive of a working computer unless there are very strong protective elements in the system. Otherwise, they are too vulnerable and too accessible.

FILE FORMATS

There are many file formats for images that are in use today, some widely and some not so widely. In the context of this book there are three that will be described in some detail because they are useful or unavoidable in forensic imaging work. A few others will be briefly described because they are widely known or have some interesting properties, but are not commonly used in forensic work. The three main formats are:

- TIFF, Tagged Image File Format (and in the PC world, .tif)
- JPEG, Joint Photographic Experts Group format (.jpg)
- Wavelet

The first one is lossless, and the second two are lossy. Lossless formats are such that the image that is reconstructed from the file is identical to the one that was compressed in the first place in all respects. Lossy formats cannot make this claim. The image that is constructed is different from the antecedent.

LOSSLESS COMPRESSION

Before getting into the file formats, however, it is useful to know of a few coding techniques that can be applied within compression processes. Two important ones are Huffman and LZW encoding. These can be applied in preparing images for several different formats.

Huffman encoding relies on the sequential nature of the pixels in an image. One can envision them as in sequential rows of numbers, or sequential columns of numbers, or one can devise other pathways through the array of numbers. Just so long as the pathway is well known and does not reuse any pixels, any one will do. The encoding process takes advantage of the fact that the items in the photo will tend to have some solid objects with continuous col-oration across some distance, say across a row. This means that there will be a high probability of finding several pixels with the same value in a continuous sequence. When such a run of pixels is encountered, one needs only to know the level that all the pixels have, where the first one is, and how many are in the run. The location of the first pixel in the run is gleaned from where in the scan across the row the run was encountered, so it is generally not necessary to explicitly mark its location. That leaves two numbers, the brightness level and the number of pixels in the run, as being all that is needed to represent the entire run. For example, if there are 20 pixels with a brightness of 154, one needs only those two numbers to reconstruct the entire run. Otherwise, 20 numbers would be needed. This simple process can achieve a 10:1

compression ratio in this simplistic example. Clearly the more irregular the image, the less compression one can achieve by this approach. On the other hand, photos with large areas of uniform content, for example blue sky, and images of documents, with lots of paper all the same color, can be greatly compressed by this simple process. Typically lossless compression techniques cannot achieve compression ratios greater than two with natural photos. In artificially generated images, such as fax transmissions, much greater compression ratios can and are achieved.

One can take an additional step and achieve an increase in compression. Typically, brightness levels in digital photography are depicted by 8 bit, binary numbers. This means that there are 2^8, or 256 possible levels. It is also true that most of the values in close proximity to each other spatially will also be close to each other in brightness. This means that the full range of 256 numbers may not be needed for a given row. One simple way to capitalize on these phenomena is to specify the brightness level of the first pixel in the row, and then use the difference between it and all subsequent pixels. The result will be that the full range of 256 values will probably not be needed in many rows, in fact the number of values needed after differencing may well be 128 or 64, or even less. Thus one does not need a full 8 bit number to represent the level for each pixel. In the case of 128, one needs only 6 bits, and in the case of 64, one needs only 4 bits. So the differencing process itself can result in some degree of compression. Combining the Huffman encoding with difference encoding usually results in an even larger amount of compression than either process alone. It is important to note that no information is lost in these processes. The recreated image is truly identical to the precompression image.

In the case of TIFF, these processes can be employed. TIFF simply requires that each element in the file is properly tagged as to how it is to be reproduced. If no compression process is used on the image, this is so indicated as well. TIFF can often achieve compression ratios of 2:1, or a bit more for certain types of images. One is not apt to achieve much more compression than this, in general.

It is now possible to advance the complexity and increase the compressibility. If one were to examine a histogram (a plot of how many pixels are at each brightness level) of an image, one would quickly see that there is a brightness level that is the most common one in the image. Moreover, there is a second most common level, and so on. In general practice, the brightness levels are assigned sequential values from 0, representing black to 255, representing white. The black to white ordering is convenient for humans, printers, and other display devices; it is not necessary for computers doing image processing. One could take the most common brightness level and assign it an alias number, say, zero, then take the second most common level and assign it an alias of 1, and so on. The approach is likened to a dictionary where the true meaning, the actual brightness level is the definition, and the alias is the word. Hence this approach is often referred to as lexicographic. The most common lexicographic process is called the LZW (Lempel, Zev, and Welch) process. The value of this approach is that the most frequently used numbers will be very low, and will not require a full 8-bit byte. The full 8-bit byte will not be needed for very many pixels at all. The result is that by simply passing along the specific dictionary used in the image, one can achieve lossless compression. And, in fact this approach can be combined with both of the methods just described. The file would be referred to as TIF-LZW.

LOSSY COMPRESSION

It is possible to achieve much higher compression ratios, but this generally requires that some information gets lost in the process. That is, the image created from the file is not a truly faithful duplicate of the image that was filed away. The reconstruction is a "copy," and not a "duplicate," in the jargon of the forensic digital imaging community. As a general rule, the amount of loss increases as the amount of compression increases.

The most common lossy compression process in use today is JPEG. With JPEG, the user can select the degree of compression desired. JPEG can be used to reduce file sizes by a factor of 80, but image quality will be significantly degraded in the process. It is often believed that this routine will lose information primarily associated with fine detail (high spatial frequency), but this is not necessarily true. Some low-frequency and color information might be lost as well. The information loss is quite complicated. This approach will insert artifacts into the image (image constructs that were not part of the original image), and again, the higher the level of compression, the more the level of artifact insertion. The information that is lost has as much to do with the artifacts as with the reduction in fine detail.

The basic premise of the JPEG process is the assertion that the world can be divided into several small segments, and that each of these can be represented by the superimposition of simple, repeating mathematical functions – in particular, cosine waves. This is an extrapolation of the Fourier theorem, in which it is shown that any repeating waveform can be replaced by an infinite number of superimposed sine and cosine waves. Mathematicians have found ways to take non-repeating waveforms and convert them into ones that repeat, thus any waveform can be represented. If one were to draw a line across an image and then note the brightness of the image at any point along the line, one would have an arbitrary waveform. Fourier says that this can be fully represented by the infinite series of sine and cosine waves. There in nothing wrong with this theory, but in practice, there is no such thing as an infinite series. So the process yields a good approximation, and the more one compresses, the bigger the gap between theory and practice.

Before describing the JPEG process it is useful to recall, again, that a digital image is essentially a string of numbers – a very long string. It might be broken down into planes (for color space) and lines (rows of pixels), and individual pixels, but eventually, it is a long string of numbers. The planes represent essentially three separate monochrome images (if the original image is a normal color image). And each plane is divided up into rows of equal numbers of pixels stacked above each other to form columns. The three monochrome images can be made to represent red, green, and blue brightness levels. Or, if working with subtractive primaries, one would use cyan, magenta, and yellow levels. There are some other, less intuitive color representations that can be used as well; each is referred to as a "color space." One color space approach in common use is first to separate out the subjective impression of brightness, and then use two other dimensions to represent the various colors possible at each brightness level. One such space is called YUV, in which the Y dimension represents brightness, and the other two add color information. These spaces are convenient mathematically, and different versions are tailored to different applications. Similar color spaces are YIQ, which is often used with color video, and LAB, which is more well tuned to perceptions, or LUV, in which changes in one dimension do not alter the perceptions in the other dimensions. Suffice it to say that it is most convenient if the initial color image is broken down into three separate, but linked packets of data before suitable formatting and compression are applied. One important feature of the color spaces just mentioned is that they separate out a subjective indication of brightness. And humans are rather sensitive to fine detail when it is captured in brightness variation, and less sensitive to details in the pure color dimensions. In fact humans are not very sensitive to slight color variation except in a few special situations. The result is that the use of one of these color spaces means that loss of information in the color dimensions is better tolerated than loss in the brightness dimension. One can take advantage of this when compressing files.

To effect compression, the overall image is divided up into a very large number of small squares of pixels, each 8×8 pixels, a total of 64 pixels per square. And, since each pixel requires three numbers for the three primary colors, this square, before compression, comprises 192, 8 bit numbers. To create the JPEG file, the image is separated into three monochrome images as indicated above. Then each of the small squares is represented by a new square. This is done to all three monochrome images. The new squares are arrays of numbers in cells, and each of the new arrays has eight cells on a side. The next step in the process is often done differently for each dimension in the space (typically YUV or YIQ). The original square has numbers that indicate how bright the image is at each of the 64 (pixel) locations. It is assumed that, since the 8×8 square is a tiny fraction of the full image, the pattern of the image that traverses the square will be very simple. The square may go from light to dark as one moves from left to right. Or it may change this way from upper left to lower right. Or it may start out light, get dark, and then get light again. But in any event, the amount of change is assumed to be quite limited and the pattern is assumed to be fairly simple. This means that the image information appears as a simple waveform traversing the small square. According to Fourier, this pattern can be adequately represented by a series of (in this case) cosine functions, each with a specific frequency and amplitude. The frequencies that will be used correspond to the cells in the second array. The upper left cell is zero frequency (or the average brightness of the full square) and each of the other cells has a frequency that corresponds to its distance from this upper left location. The highest frequency cell is in the lower right cell. Instead of an infinite series, there will be only eight frequencies, but the resolution limit of the original image is dictated by the number of pixels it has, and this is not greater than the frequencies available. So frequency loss is not really an issue at this point in the process. The number that goes into each cell represents the amplitude of the cosine wave at that discrete frequency. This is calculated from the original according to well-known formulae. There are seven cells to the right of the initial cell, and seven down the rest of the first column. The numbers in these cells deal with vertical and horizontal components of the original image waveform. The other cells emanate from the initial cell at a variety of angles. The entries in these cells correspond to a fixed set of angles between horizontal and vertical that the waveform might take. The only loss incurred so far is that the estimation of the amplitudes is limited by the use of only 256 integer numbers (there will be round-off errors) and the angles at which waveforms can traverse the original square are limited by the geometry of the square array – not all angles can be obtained in an 8×8 array. These errors are miniscule, and if the image file at this point were to be reconstructed, the amount of error may not be detectable.

Compression is implemented via the next two distinct stages. The first is "quantization" and the second involves the Huffman, differencing and lexicographic process mentioned earlier. To start the quantization process, the operator must choose a level of compression. This will set up a quantization, or round-off table. Originally, all of the numbers in the array of cells are 8-bit, integer numbers. That is, they are each one of the integers between zero and 255. Remember they indicate the amplitude of the cosine waves at the angle and frequency designated by the cells' locations in the array. If the operator wants only a minimal amount of compression, the quantization might be set to, say 7 bits (instead of 8). This means that instead of 256 numbers, there are now only 128. Each of the original numbers was mapped into a new number – basically a rounding off process. So, zero will now represent both zero and 1 from the original set. Two and three will become 1, and so on. Once these conversions are made, it is no longer possible to convert back and create a true duplicate! Also notice that since there are fewer numbers available in the set, more of them are the same. If a large amount of compression is wanted, the new set of numbers can be the 3-bit set – which contains only eight numbers! The quantization process tends to make the numbers in the various cells more closely the same as their neighbors. In many cases, the quantization can be skewed so as to impose greater rounding off on the higher frequency numbers (the ones in the cells to the right and below the initial cell in the upper left hand corner). This will make most of the numbers in the lower right corner zero.

Once the quantization is complete, Huffman, differencing, and lexicographic compression can be employed. But in JPEG compression, instead of going across rows, a snake-like sequence is followed. Consider the head of the snake to be in the upper right-hand cell, and the body then extending in sequential diagonals through the rest of the array. The tail comes out in the lower right-hand cell. As it turns out, this arrangement tends to maximize the probability that adjacent cells along the sequence are of the same value. This means that the whole sequence of 64 numbers might be reduced to only a handful. The end result is that the more drastic the quantization, the smaller the number of numbers. Any cells that represented image squares with very little image detail might be reduced to a single number. Note that compression is possible in two ways. If the original numbers are 8-bit, and the quantization brings this down to, say 4, then it is now possible to have two 4-bit numbers per 8-bit byte. Also, when the application of the run-related compression processes is completed, the actual number of numbers needed can be quite small indeed. Finally, the level of compression of the color-related dimensions is usually more than for the brightness dimension. This leads to even greater compression.

What is lost in the process is the ability of the system to estimate the image values from the quantized cosine waves with accuracy, the ability to reproduce angles accurately, and since all brightness and color components are processed, coordination among the colors in the final image is at risk. If there is a high degree of spatial frequency information in the original image, and it is pronounced, it will be present in the reconstructed image as well. However, it may not be in proper proportion to other information due to quantization. And, if the image comprises swirling, fine lines, each of these may well be broken down into a sequence of straight-line segments. In places where there are dark areas that abut light areas, there may be highlighting in which the dark portion is made even darker near the edge and the lighter portion is made lighter just inside its domain. In areas where there are relatively low frequency changes in color, the colors may reproduce differently, changing the appearance of the image by creating small, artificial bits of color that are added capriciously. Finally, on close inspection, or if blow-ups are made, the 8×8 pixel squares will be visible in the image.

As a practical matter, in normal scenes, and where the compression levels are moderate, less than five, these artifacts are minimal. It is only when the compression ratios approach 30 or so, that the artifacts become a significant issue. So, in normal photographic images, there is not likely to be a problem. But, if the images are of specialized objects, such as fingerprints and shoe impressions, it is not possible to give a general rule regarding the use of JPEG compression other than to say avoid it if possible and use minimal compression when it must be used (because, for example, the camera automatically invokes JPEG compression).

A new version of JPEG is due out very soon. It is to be called JPEG 2000 and it is expected to result in much less damage to images at the same compression levels. At this point in time, it is too soon to comment, other than to say, it will probably be a welcome improvement.

Because the JPEG 2000 specifications are still under development, this document will not go into the details – just be aware that it is in the future.

As it turns out, JPEG is essentially a specialized case of so-called wavelet compression. The cosine is the waveform used in JPEG, whereas other waveforms are used in wavelet compression. By carefully selecting the waveform to be used according to the nature of the image to be encoded, it is possible to obtain lower losses at the same levels of compression. For example, it may well be the case that a particular waveform is much better at compressing fingerprint images than the discrete cosine function used in normal JPEG. This does not mean that the particular waveform selected will outperform JPEG when applied to typical snapshots. So, while wavelet compression is widely used, it is not in wide use in forensic digital imaging at this point in time.

FILE MANAGEMENT

Aside from issues of tampering, the important issues in managing current image files are assuring the ability to access those needed at any point in time and assuring that the images retrieved are the ones sought. Obviously the keys to this process are attaching descriptive information to image files in such a way that the information stays attached to the given image, assuring that the attached information is unique, and that the attached information can be searched in a logical way. The information which is attached to an image is often referred to as metadata, and the way to assure that the files can be searched is to build a database out of the metadata and insert pointers that will lead to the indicated image file. The image files can also be other types of files, such as documents, graphs, video clips, sound clips, and so on. As a class, these items are referred to as "assets." The simplest form of metadata is keywords. A series of single words, numbers, or small groupings of words can also be used. Additionally, it is possible to attach paragraphs, which can be searched via a keyword search, or a more complex phrase and/or sentence parsing approach. In any event the words and numbers (which includes dates) are organized into a structured array that can be easily searched. For example, the word "May" will mean the fifth month of the year if it is in the date column, or it will be a person's name if it is in the name column. In this way the actual word can take on meaning by virtue of how it is placed in the database. There are many techniques used in database technology to facilitate a search on dates, or names, etc.

There are a number of software packages currently on the market that do this, but none fully meets the needs of the forensic community. Some will not work with a full set of asset types, for example they will only deal with text documents or line images. Some have significant limitations on how many assets can be managed, and others have cumbersome search procedures. At one point there was a rather good offering, but it was a bit ahead of its time, and did not survive in the marketplace. There are new products appearing on the market, but care should be taken to use a very popular or fundamental approach to selecting file management software. If a large database is built using a particular software package, and the vendor leaves the market, the package will not be upgraded as operating systems evolve. The result may well be that at some point the database and its contents will be lost. The practitioner will need to find a suitable multimedia database software package to file and retrieve assets, or before very long, he or she will have a write only memory. The files are large, there are many of them, and one cannot thumb through them very easily, as one can do with a paper file. The important message is to structure the filing process early on and update it continuously, before the burden of finding older items becomes insurmountable.

LEGACY FILE MANAGEMENT

Most courts require that information used at trial be available for long periods of time after the case is heard. In many instances the time can be in excess of 50 years. This is a virtual eternity in a field in which new products enter the market every six months and most products become so out of date as to be unusable in just five years. Nonetheless, the requirement is there. The assets must be available, technology turnover notwithstanding. This begs the question of legacy file management.

In just three decades, punched paper tape gave way to reel-to-reel magnetic

METHODIST COLLEGE LIBRARY
FAYETTEVILLE, NC

tape, which gave way to magnetic disks, which gave way to optical disks, and this will not stop any time soon, and there are at least two decades to go to reach the 50-year requirement.

The problem is several levels deep. Having a punched paper tape in a file today is essentially a total waste. There is no device to read it anywhere close by. And, even if there were, there is probably no software driver that today's computers can use to control the reader. Finally, the file format has probably gone into oblivion years ago and the software that was written to work with the data on the tape is no longer available, and even the operating system on which these software packages depend has disappeared.

To deal with this unavoidable problem, one must assure that the agency is prepared to read and rewrite old files on an ongoing basis, upgrading the file formats to the newer standards, replacing media that are being phased out in favor of new media, and in general, faithfully migrating the stored information to new platforms as the older ones appear on the edge of becoming out of practice across the industry. Failure to keep ahead of the trends can place information totally out of reach.

Typically the failure mode is like the following. A piece of equipment fails. A search for a repair agency turns up no viable options. So the device has to be replaced. A newer version is acquired, and the drivers are installed. All of the older devices are also replaced in an upgrade program. Then the host computer is outfitted with a new version of the operating system. Current files are fine, but when an attempt is made to access an older file, one learns that the new device cannot read the older records, and the new operating system precludes running the drivers needed for the old style devices. The material, for all intents and purposes, is lost. Vigilance is the answer. There are service bureaus that retain somewhat older equipment and can make transcription for a fee. They and other consultants can also monitor the marketplace and advise on when to make upgrades. These services can be obtained if the capability does not exist in-house, but it is wise to assure that some process is put in place to protect older records.

CHAPTER 5

DIGITAL IMAGING WORKSTATION

INTRODUCTION

If you ask seven different people what a digital imaging workstation consists of, you will get at least seven different answers. This is because there are very few standards established on what should be included in the package and because many users have their own specific preferences and workload. Companies selling digital imaging workstations as a system usually establish their own standards. Others will design to meet your needs. It should be common sense that the design is going to vary; the hardware and software needed on a system that is going to be used at a prepress business is not necessarily going to be the same as what you would use in a forensic lab to analyze fingerprints.

Some of the basic features will be the same, but the digital imaging workstation should be designed from the ground up to fit the needs of the user. In this chapter we will discuss the basic requirements of a system for use in a forensic lab. We will start with the case that holds everything and then discuss the other components, such as the motherboard, CPU, RAM, hard drives, video card, monitor, and why we would choose each for our digital imaging workstation. We will also discuss input/output devices and software needed to operate the digital workstation.

A basic digital imaging workstation should contain the following items.

HARDWARE

- Computer system, which will include the computer, monitor, keyboard, and mouse
- CD-ROM drive
- CD-writable drive
- PCMCIA card reader
- Iomega Zip drive
- Flatbed scanner
- Negative scanner

- Video capture
- Photo-realistic printer

SOFTWARE

- Image enhancement
- Analytical
- Word processing
- Presentation
- Video capture
- Panorama
- Image database
- CD authoring
- System diagnostic
- Spreadsheet

There are usually three things that buyers consider right up front when shopping for a digital imaging workstation. Number one is usually the cost (does it fit the budget), number two is performance (will it do the job intended), and number three is ease of use (is it user friendly?). Some may even swap the priority of two and three. With these three factors in mind, let's consider the digital imaging workstation.

COMPUTER CONFIGURATION

Start with the case where most of the other components are mounted and stored. One might not think this is a big issue, but there's a purpose for each case design and why it is used. Most end users would choose computer cases by outside appearances and prices. Most computer experts would choose by inside functionality and brand. It is a good idea for you to consider all of these factors.

CASE AND POWER SUPPLY

A very critical part of a computer case is the power supply. The power supply is one of a few PC components that have any moving parts (cooling fan). The wattage of the power supply is the important factor: 230 W is more than enough for the majority of desktop applications, but at least a 300 W unit will be needed for a digital imaging station. This is because it will be running multiple hard drives and CD drives, which require more power. It is recommended that one stay with major brands and purchase from a reliable source.

Next, a case fan is highly recommended! The failure of a power supply or

CPU cooling fan can go unnoticed for months, causing significant temperature rise inside your computer case. High temperature is one of the worst enemies of all computer components. A fan blowing hot air out of the front of a case, combined with a power supply fan that sucks fresh air in, helps maintain a relatively low constant temperature inside the case. The other benefit of blowing air out of the front of a case is to avoid accumulation of dust on the ventilation holes in the front of a case. Take a look at the back of an old power supply and you'll see large accumulations of dust, blocking the air flow.

The most popular industry standard among case form factors is the ATX. Case form factors determine how the case is built and what motherboards it will accommodate. As an example, the ATX case will only use an ATX motherboard. The main advantage of ATX is its advanced power management, which allows a system power-down without pushing the power button. In addition, there is an extra PS2 port for the mouse that saves a COM port for other purposes. In comparison, the old AT form factor is being phased out slowly. The MicroATX is a smaller version of ATX with a significantly reduced number of expansion slots (and thus reduced functionality of a system). The MicroATX was designed to save cost, but has not yet become popular, mainly because of its reduced functionality.

MOTHERBOARD

The motherboard is something on which you do *not* want to "save money." It is not easy to replace a motherboard, even for professional technicians. Besides, if the motherboard is bad, your whole system is likely in jeopardy.

The reliability of the motherboard is measured by return rates and is roughly correlated with the price you pay. As a rule of thumb, the higher the price from a reputable supplier, the better the reliability. We suspect that manufacturers with reputations to protect are more likely to do more extensive testing and quality control before shipping. Motherboards have no moving or consumable elements. If they are going to go bad, most often they do so within a month. Therefore, we suggest shop carefully for the motherboard and get the best your budget can afford.

Make sure your motherboard form factor (ATX, AT, or Micro ATX) matches that of your case. Buy the latest technology! Yesterday's technology means more difficulty or limitation in future upgrades. For example, motherboards only support certain speed CPUs and if later you want to upgrade to a faster machine, you would have to change the entire motherboard.

Last, but not least, make sure your motherboard is compatible with your microprocessor. There are two main factors to consider: processor form factor and bus speed. Processor form factor relates to the actual socket in to which it will mount. Examples include Super Socket 7, Slot 1, or PPGA. Bus speed relates to the actual speed of the processor, such as 400 MHz and so forth. Choose your motherboard carefully.

Another thing you will encounter when choosing motherboards is integration. This seems to be the current industry trend. It saves cost and space to integrate as many components onto the motherboard as possible, such as the video, audio, modem, and network cards. Integrated motherboards tend to have limitations on future upgradability and expandability. It is not recommended for the digital imaging workstation.

CPU (MICROPROCESSOR)

The choice of microprocessor for your digital imaging workstation is down to mainly three: Pentium (including Pentium II, III, and IV), Celeron, and AMD.

Besides the obvious price differences, the main technical difference between Pentium II/III and Celeron is that Pentium II/III has 512k level 2 cache, while Celeron has only 128k.

Cache is used to speed up the CPU access to repeatedly used files/data in memory and it is very critical for activities such as gaming, 3D computing and graphic work. If you only plan to use the computer for word processing, e-mail, and surfing the web, Celeron is a very good choice. However, for the digital imaging workstation, where much of the work is with graphics, the Pentium II/III/IV with more cache would be a better choice.

AMD microprocessors have gained popularity recently. Their price, speed, and 3D performance are a very attractive alternative. AMD processors tend to generate more heat than the Intel counterpart, therefore a better heat sink and/or cooling fan is recommended. A case or system fan would be a sound investment too.

Among all computer components, microprocessors are among those with the least return rates and thus the highest reliability, which is true for all brands of microprocessors.

You should choose the highest speed your budget will allow. Keep in mind that the overall speed of a computer system depends on: microprocessor speed, size of RAM, size of cache, hard drive access speed, and video speed.

Also keep in mind that different microprocessors may require different motherboards. Ensure that the motherboard and the microprocessor are compatible.

RAM (RANDOM ACCESS MEMORY)

Memory is one thing that we suggest you do not buy just based on prices alone. The quality of memory modules is extremely critical to computers. This is especially true in higher-speed, 133 MHz and 100 MHz systems. That "blue screen" with a "fatal error" message is often a sign of memory problems.

Try to use modules only with the name-brand chipsets that carry the names of the original chip manufacturer. Name-brand chipsets include Micron Technology (MT), Texas Instruments (TI), Toshiba, Hyundai (HY), Goldstar (LGS), Samsung (SEC), Siemens, NEC, Panasonic, and Fujitsu.

Adding more memory modules is probably the easiest and sometimes most inexpensive way to upgrade an existing system.

Seventy-two-pin EDO memory modules are 16-bit technology, but Pentium systems that use EDO memory are 32 bit. Therefore, EDO memory must be used in pairs so that two banks of memory are combined to allow 32-bit processing. Unless you already have a piece of EDO memory, we suggest you purchase EDO memory in pairs for upgrade purposes. Because they have to work together, we suggest using only identical pairs.

HARD DRIVES

Hard drives are usually divided into two categories, SCSI or EIDE, and the basic difference between the two is access speed (how fast can the information be accessed). SCSI (Small Computer System Interface) requires that an extra controller be installed on the motherboard. EIDE, Enhanced Intelligent Drive Electronics or Enhanced Integrated Drive Electronics depending on who you ask, is the controller built into all motherboards.

Ultra DMA EIDE is a fast technology, with an average access time of 11–12 ms. In comparison, SCSI II (often also referred to as Ultra or narrow SCSI) hard drives typically have an average access time of 9.5 ms. In our opinion it is not worth the extra money to go with SCSI II. Ultra DMA is just fine for most applications.

If you are serious about access speed for running databases and servers, consider using Ultra-wide SCSI III (often referred to as "wide"), which have a typical access time of 7.1–7.5 ms. Then, if you are *really* serious about getting fast access to your data such as large databases and image files, consider using one of the latest hard drive technologies, Ultra-2-wide SCSI III with an access time of 5.2–5.7 ms. Understand, you'll pay dearly for the speed. SCSI hard drives are more expensive and require that a separate SCSI controller be purchased and installed.

In most cases we think the Ultra DMA EIDE hard drive is capable of handling any function that you will accomplish on the digital imaging workstation. There are cases however, such as servers that might be used to store images in a database and need quick access that SCSI hard drives would be better.

There is not much difference among the name-branded hard drives. Some buyers prefer one brand to another, because of historic and one-time bad experience. Yet, some others may just have the exact opposite brand preference. Therefore, we suggest going with the highest capacity (gigabytes) for your buck in choosing hard drives.

Do not buy refurbished hard drives no matter how low the price. The potential trouble down the road is not worth the saving. Used, but working, hard drives are a better and possibly more reliable way to save money.

VIDEO CARD

Most people do not work with graphic subjects and tend to underestimate the need for a good video card. The major consideration factors in choosing a video card are video speed, resolution, and price. Consider these relative to the four categories of PC users: average, above average, graphics workers, and gamers. For average users who use computers only for word processing, spreadsheet, and e-mail, an inexpensive 4–8 MB video card would do just fine. For them, price may be the most important factor. For above-average users who surf the web on a daily basis in addition to other common PC tasks, an 8 MB video is a must and 16 MB may be optimal. Modest speed and price should be the main factors for consideration. For graphic designers and CAD workers, resolution is the most critical factor. Higher resolution typically means slower speed. Therefore a video card with a combination of high-resolution chipset and high memory capacity is recommended. Keep in mind that the highest resolution one can achieve is also determined by the maximum monitor resolution. That's why graphic designers often use high-resolution, low-dot-pitch professional monitors.

Video card speed is determined mainly by three factors: chipset speed, number of onboard memory chips, and the type of onboard memory. SGRAM is faster than SDRAM. If you care about speed, make sure you get the faster memory chips on the video card.

If you have an old computer and want to upgrade it for faster video, first determine if you have an AGP slot. If not, your choice would be limited to PCI video cards, which are slower and being phased out. High-end PCI video cards are hard to find.

MONITORS

Size is more important than dot pitch for most users. Once you have used a 17" you don't want to go back to a 14" any more. Buy the largest size your budget will allow. The price differences among 14", 15", and 17" models have shrunk significantly in the past two years.

Here is another consideration for buying big. The monitor is almost the only big-ticket item in a computer system that will last for more than three years without losing the technical edge and its total value. Everything else seems to lose value significantly within a year or so. Therefore, spend a little more on a monitor – it is a sound investment strategy.

For professional graphics, image, and design work, you need the highest resolution your money can buy. A "0.27" dot pitch is a must. A 0.26 or 0.25 is even better. The dot pitch indicates the size of the pixels on your monitor screen measured in millimeters. The smaller the dot pitch, the finer your image, and thus the better.

Do not choose monitors with integrated speakers. Yes, the built-in speakers would save you desk space, but may not necessarily save you trouble. Most computer speakers are much cheaper than monitors and tend to have higher failure rate than monitors. It would be too much of a trouble to send a monitor back to your vendor just because of defective speakers.

A monitor displaying 640 × 480 means it can display 640 vertical lines and 480 horizontal lines. A monitor display at a higher resolution can display more information and finer graphics. For example, a user will find that more columns can appear on screen in an Excel spreadsheet when the monitor runs at a higher resolution. Table 5.1 shows recommended minimum resolutions for different screen sizes.

Monitor size	Comfortable resolution
14"	800 x 600
15"	800 x 600
17"	1024 x 768
19"	1280 x 1024

Table 5.1

Recommended minimum resolutions for different screen sizes.

INPUT DEVICES

Common input devices such as keyboards and mice are something you will most likely use every time you turn on your computer. These devices tend to have mechanical and moving parts that could cause reliability problems. Therefore, it is a good idea to choose brands and models with good reliability and comfort.

Ergonomic keyboards are highly recommended. You won't feel thrilled when you first use one, but getting used to it is easy and does not take too long. After a few days your wrists will enjoy it more than you think.

Intermediate storage devices (Zip Drive)

Understand that "intermediate" is just what it means, this is not a medium to which you archive your images. It is used as an intermediate storage device prior to writing the images to an archival medium, such as a CD-ROM.

Zip drives are available in two capacities, 100 MB and 250 MB. The 250 MB drive can read the 100 MB cartridges, but the 100 MB drive cannot read the 250 MB cartridges.

Internal and external models are both available. The internal models will be either SCSI or IDE drives. External models are available with SCSI, USB or parallel connections.

The Zip cartridges can be removed from the drive, write protected and placed in a secure area.

A Zip cartridge can be write protected easily by using the software that came with the drive. Insert the cartridge that you want to protect into the drive. Open "My Computer" and right-click on the "Zip" icon. The shortcut menu will open.

Select the "Protect" command. The "Disk Protect" options dialog box will open.

Two options will be available:

- Write Protect – others can read the disk, but they cannot change any of the information on the disk.
- Read/Write Protect – the disk cannot be opened; information on the disk is protected, both read and write.

Read/Write Protect is highly recommended if these are your primary images and they have not yet been written to CD-ROM (see Chapter 8).

Notice that when you read/write protect a Zip cartridge, if the password is lost the data on this disk can never be recovered. It must be reformatted. So be careful!

Any primary images on a Zip cartridge should be written to a CD-ROM as soon as possible.

CD-ROM (Compact Disk Read Only Memory)

A CD-ROM is a type of optical disk capable of storing large amounts of data – up to 680 MB per disc. A single CD-ROM has the storage capacity of 470 floppy disks. CD-ROMs require a special machine to record the data, and once recorded, they cannot be erased and filled with new data. To read a CD, you need a CD-ROM drive. Almost all CD-ROMs conform to a standard size and format, so it is usually possible to load any type of CD into any ROM drive. In addition, most CD-ROM drives are capable of playing audio CDs, which share the same technology. CD-ROMs are particularly well-suited to information that requires large storage capacity. This includes color graphics, sound, and especially video.

A CD-ROM drive is a device that can read information from a compact disc. CD-ROM drives can be either internal, in which case they fit into a drive bay and are connected to the motherboard through an IDE onboard controller or SCSI connected to a separate interface card, or external, in which case they are usually connected to the computer via a parallel port, SCSI port, or USB. Parallel CD-ROM drives are easier to install, but they have several disadvantages. They are more expensive than internal drives; they use up the parallel port, which means that you can't use that port for another device such as a printer; and the parallel port is usually not fast enough to handle all the data pouring through it. SCSI CD-ROM needs an extra SCSI controller board to connect it to the system. Typically, SCSI controller boards are more expensive than IDE ports. USB has become somewhat the standard for external CD-ROMs. They are user friendly and can be hot-swapped from system to system.

There are a number of features by which CD-ROM drives are classified, the most important of which is the speed. CD-ROM drives are generally classified as single-speed, double-speed (2×), triple-speed (3×), quadruple-speed (4×), hex-speed (6×), odo-speed (8×), deca-speed (10×), triple quad-speed (12×), or higher.

Within these groups, however, there is some variation. Two more precise measurements are the drive access time and data transfer rate. The seek time, also called the access time, measures how long, on average, it takes the drive to access a particular piece of information on a disc. The data transfer rate measures how much data can be read and sent to the computer in a second. Aside from speed, another feature by which CD-ROM drives are classified is compatibility with existing standards. If you plan to run CD-ROMs in a Windows environment, you will need a drive that conforms to the MPC II standard.

We must mention the DVD drive, short for Digital Versatile Disk, because it is the newest standard for optical storage. This specifies a disk that is the same size as a standard CD but is able to hold much more information. DVD was developed by an industry consortium of electronics companies led by Toshiba and Philips and it is anticipated the new standard will usher in a new era of growth for multimedia, interactive applications on the PC. For PC applications, DVD will be used the same way the CD-ROM is used today. The main attraction for PC users is the larger capacity of the DVD disk. The capacity will be 17 GB on a double-sided dual-layer disk, 8 GB on a single-sided dual-layer, and 8.4 GB on a CD-R which will only be single layer but could be double- or single-sided. The PC will also be able to play DVD-Audio and DVD-Video, two standards developed primarily for home electronics devices. New ROM drives, called DVD-ROM, have arrived on the market to accommodate the new disk format.

Advances in digital video and audio standards are coinciding with the development of the DVD to create unprecedented multimedia storage capacities. For instance, MPEG-2 video and audio compression will be used to compress still images such as scanned or pictures as well as full motion video. This compression will allow much greater storage than you could attain using a hard drive; for instance, a single-layer, single-sided DVD has enough capacity to hold 2 h 13 min of video. Some DVD-ROM drives will have a built-in MPEG-2 chip, and some others will use an MPEG-2 chip installed in the PC.

CD-R/RW (writable/rewritable)

CD-RW is a rewritable CD utilizing an innovative phase-change recording material. This breakthrough effectively increases data erasability and recording sensibility, which are two major drawbacks of conventional phase-change recording material. During the write operation to a CD-RW, recording switches between the "amorphous" phase (recorded state) with lower reflectivity and the "crystal" phase (erased state) with higher reflectivity. Areas on the surface are recorded or erased according to the temperature rising/falling through laser beam irradiation. The CD-RW is ideal for personal uses, such as data back-up and making original CDs, because it allows users to overwrite repeatedly.

CD-R is a write once disk, which means the information written to the disk cannot be overwritten. It uses a laser technology and actually burns holes in the disk as it writes. With CD-R, users can make disks for distribution, create electronic albums of digital camera images, and store data that should not be revised or deleted. This is the medium that is recommended for storing and archiving images.

A CD writable/rewritable drive is a device that can write information to a compact disk. CD-R/RW drives can be either internal, in which case they fit into a drive bay and are connected to the motherboard through an IDE onboard controller or SCSI connected to a separate interface, or external, in which case they are usually connected to the computer via a USB port, parallel port, or SCSI port. Parallel CD-R/RW drives are easier to install, but they have several disadvantages. They are more expensive than internal drives; they use up the parallel port, which means that you can't use that port for another device such as a printer; and the parallel port is usually not fast enough to handle all the data pouring through it. SCSI CD-RRW needs an extra SCSI controller board to connect it to the system. Typically, SCSI controller boards are more expensive than IDE ports. USB has become somewhat the standard for rewritable drives.

REMOVABLE MEDIA

PCMCIA cards

Personal Computer Memory Card International Association reader is a long name for a device that can read the cards that store the images in your digital camera. What type of flash card reader you install will depend on the cameras you are purchasing. They may be called PCMCIA readers, compact flash card readers, or digital film readers. Card readers feature high-speed data transfer rate which is at least 20 times faster than traditional serial port connectivity and even faster than a USB connection.

Memory cards for digital cameras come in three types, Type I, Type II, and Type III. There are basically two differences, physical size of the card itself and its capacity.

Type I cards

Type I is usually referred to as a compact flash memory card. It is the super-small, reliable, removable storage solution for current and next-generation consumer electronics and digital cameras. Weighing only half an ounce, the compact flash card is one-fourth the size of a standard PCMCIA card. Its compact size, ruggedness, single power supply, and low power requirements, make the compact flash card ideal for reliable data storage in digital cameras. Compact flash cards can mount independently as a 50-pin cartridge, or be easily slipped into a 68-pin Type II adapter for compatibility with any existing PC Card Type II or Type III slot. Storage capacity ranges from 4 MB to 128 MB.

Type II cards

Type II is usually referred to as an ATA flash PC card and has the same width and height as the Type I, but is about twice as thick. Its storage capacity ranges from 4 MB to 160 MB.

Type III cards

Type III are usually referred to as PC card hard drives. They offer larger data storage capacity. These are actual hard drives similar to the ones used in early laptops. They are normally used in the higher-end digital cameras, but are easily damaged. The PC card hard drives are an excellent solution for primary as well as secondary data storage in 520 MB and 1040 MB capacity. Most Type III PC card hard drives utilize the PCMCIA standard so you can interchange them between notebooks and desktop computers with Type III slots.

Smartmedia

Smartmedia is another removable medium on the market. It looks similar to the compact flash card, but is much thinner. The packaging meets the stringent criteria which these types of applications demand including low cost, durability, light weight, and small footprint.

Smartmedia cards allow fast and easy data transfer between digital camera and desktop computer. Smartmedia is currently used in the Solid State Floppy Disk Card (SSFDC) removable storage media for handheld consumer electronics products such as digital cameras.

You must use the optional card adapter to read the cards in any Type II or III ATA compatible PC card slot. This lets users read the cards in notebooks, Type II digital cameras, and desktop computers equipped with a PCMCIA card reader.

Another removable medium on the market now is the memory stick, which is proprietary to the Sony camera. An adapter must also be used to read the memory stick in any standard card reader. Sony has also released a camera that writes directly to a compact disk. These disks can be read in any standard CD-ROM.

Readers that can read all cards are available with IDE interface (internal) which would connect to an existing IDE controller on the motherboard. They are also available with a SCSI interface and PCI interface, both of which require the installation of a separate controller card. Inexpensive card readers are available (external) that connect to the parallel or USB ports. These readers usually only accept the Type I and Type II cards, along with the smartmedia and memory stick if you have the adapters.

The card reader with the IDE interface is sufficient for most uses on the digital imaging workstation. It will read all three types of cards and the smartmedia and memory stick with the correct adapters. It is less expensive than the SCSI/PCI readers and requires no extra controller cards to be installed.

Image storage devices for digital cameras are a technology that is rapidly changing. At present, compact flash cards are the most popular. They are rugged, easy to use and very widely available. All other approaches have some limitations.

SCANNERS

Two scanners are necessary for the digital imaging workstation, a film scanner and a flatbed scanner. When choosing either scanner look for the same features, usually resolution and dynamic range.

What is dynamic range? Image density is measured from image brightness with optical densitometers, and ranges from 0 to 4, where 0 is pure white and 4 is nearly pure black. Less brightness is more density. Density is measured on a logarithmic scale like the Richter scale, and density of 3.0 is 10 times greater than a value of 2.0. An intensity range of 100:1 is a density range of 2.0, and 1000:1 is a range of 3.0.

Virtually nothing you will scan will be greater than 4.0. The extreme values of density capable of being captured by a specific scanner are called DMax and DMin. If the scanner's DMin were 0.1 and DMax were 3.2, its dynamic range would be 3.1. Greater dynamic range can show greater image detail in both the dark shadow areas of the photographic image and the brightness in the same image. Basically the range is extended at the black end of the image.

When we say the "black end," we speak of positives, either prints or slides. When images from negatives are reversed, this effect transfers to the highlight tones. Most literature describing 30 bits just says it improves "shadows and high-lights" without making this distinction.

A printed magazine image probably has a dynamic range well less than 2.0. Photographic color prints might have a dynamic range of about 2.0. Film negatives might have a range up near 2.8. Slides may go well over 3.2. These are not precise numbers. Twenty-four-bit scanners might have a dynamic range of near 2.5, 30-bit scanners might be 3.0, and 36-bit scanners might be 3.4. Only rotating drum scanners can approach 4.0 (Photomultiplier tubes, PMTs, extremely expensive). And of course, all scanners are not equal; some will have higher dynamic range than others because they have less noise.

The dynamic range of a 30 or 36-bit scanner can sometimes capture more detail from an image than can a 24-bit scanner. Specifically, it can add detail in the shadow tones of prints and slides, and in the highlights of images from negatives. It usually will not be a major consideration for scanning photo prints. It can make a very noticeable difference in scanning film, both slides and negatives.

While there are exceptions, typically 300 dpi flatbed scanners use 24-bit color, and 600 dpi scanners use 30-bit color. Typically film scanners are 30 or 36 bits. And more bits imply more dynamic range. It is not a given, however; quality is also a large factor.

Why is 30 bits better? Twenty-four-bit color is three 8-bit bytes, one each for red, green, and blue, to describe the color of each pixel in the image. Thirty-bit color uses 10 bits for each of the three primary colors. In binary numbers, each bit is a power of 2, meaning that the ninth and tenth bits can hold four times larger numbers ($256 \times 2 = 512$, and $512 \times 2 = 1024$). So 10 bits can hold 1024 unique numbers ranging from 0 to 1023, and 8 bits can hold 256 numbers from 0 to 255. The 30-bit scanner divides the scanned density range into smaller steps, 1024 steps instead of 256 steps, and therefore can show slightly more detail in the shadow areas, for a couple of reasons. Tiny variations that might be the same color at 24 bits could be four slightly different shades at 30 bits. Tiny differences, and only at the black end, but that's more detail. And the possibility of four times larger numbers provides an opportunity of extending the dynamic range a little way into the next "10 times" logarithmic density interval. With the greater dynamic range, 30-bit scanners can deliver slightly better detail in the shadow tones of images.

Only a few top-end scanners can return more than 24 bits to the computer. The popular scanners we can typically consider (and afford) only return 24 bits, regardless of whether they are 30 or 36 bits internally, and most graphics programs or file formats don't handle more than 24 bits either. Frankly, that's not important. The main point of this extra range internally within the scanner is that it allows the 30-bit scanner to capture a higher range of image density for its internal gamma processing.

What is resolution? This is the dots per inch (dpi) the scanner reproduces as it scans the image. It seems even the scanner makers are now caught up in the more-is-better routine. A few years ago 300 dpi, 16 shades of gray was an incredible scanner. Now it seems if you don't have a 600 dpi 24-bit scanner you're likely to get laughed at by everybody in the scanning world. But do you really need 600 dpi?

A flatbed scanner is merely a series of CCDs (charge-coupled devices, light-sensitive integrated circuits) mounted in a stationary row that light reflected from a piece of flat art is allowed to pass over. These CCDs register presence or absence of light (On/Off), thus producing a pixel electronically. Since they are mounted in a single row, that is the way the electronic file is created, row by row. Essentially the CCDs are reflected one row of the flat art at a time until the image is completely built.

That being the case, resolution, or the number of pixels written based on what is reflected, is controlled two ways. The number of pixels horizontally is controlled by how closely the CCDs are placed next to each other along the single row. The number of pixels vertically is controlled by how slowly the light bar and mirror inch along the length of the flat art thus reflecting onto the CCDs. Therefore, the more CCDs and the smaller the steps of the advancing light bar the greater the resolution.

There is a physical limit to how many CCDs can be placed side-by-side in a scanner and this is known as the scanner's maximum optical resolution. This physical limitation has been breached by what is known as interpolation. Interpolation is a software/firmware process whereby the scanner essentially samples two pixels and averages (oftentimes using more complex formulas) the two pixels together to form an extra pixel (or more) in the middle. Better scanners now do this in hardware, but some still rely on their scanning software to do it. But nevertheless, this higher resolution is only psuedo-data. That is, it is data being created by averaging and not by actually sampling it from the original art.

Another interesting development is that some scanner manufacturers are indicating their resolutions in non-uniform terms. For instance, Microtek currently indicate that their scanners are 600×1200 dpi. While this seems like a higher resolution scanner than one that is merely 600×600 dpi, think about it for a moment. This measurement reflects how much data that the scanner can acquire in a square inch, or $X \times Y$. What would happen if we acquired 600 spots in the X dimension and 1200 spots in the Y? Either we wouldn't have a square, we would have noticeable gaps in one dimension, or the most likely scenario, there would be overlapping spots in one dimension. Scanners that have non-uniform resolutions don't actually give you the ability to acquire image data at this non-uniform resolution, they instead interpolate one dimension. At 600×600 they interpolate the 1200 dpi dimension down to 600 dpi (usually done by merely running the stepper motor that moves the light bar at twice its minimum rate), or at 1200×1200 they interpolate the X dimension.

So the question becomes, why scan so high if the data won't be used? There is a formula for this of course: scan at 1.5 times the lines per inch (lpi) of the final output device. Therefore if you are outputting to a 2400 dpi imagesetter at 150 lpi, then the normal maximum resolution you need to scan at is only 225 dpi. So if the uninformed user scanned his photo at 600 dpi thinking he needed that high-resolution capability because he was going out to a high-resolution imagesetter, he would be sending over nine times too much data to the imagesetter. This would result in a very long RIP time and possible crash of the RIP. (RIP is an acronym for Raster Image Processor, the device found inside a laser printer or laser imagesetter where electronic signals are controlled. The RIP is made up of software and hardware whose job it is to interpret the signal generated and sent by the computer and then form a pattern of dots so the printer's recorder can draw it on paper using a laser beam.)

High-resolution color becomes a different story somewhat. It is possible to find a continuous tone color output device where it would be nice to output a true continuous tone modification that would rival the original (i.e., National Enquirer Photoshop modification of a 35 mm slide at high resolution, which is then re-output to 4×5 inch negative on a film recorder to produce a retouched print). However, continuous tone output requires very high resolution to produce satisfactory results. This requirement pushes the upper envelope of flatbed scanning (800, 1200 dpi) and becomes a job for the slide scanner and drum scanner. Unfortunately, at these ultra-high resolutions the personal computer becomes a liability. One of us has worked on a photo retouch of a 35 mm slide that we intend to re-output to 4×5 inches on a Solitaire film recorder – the full resolution file (4800 dpi) was over 250 MB! This is definitely not a job for Photoshop.

DIAGNOSTICS

A typical method of running daily diagnostics is described below.

- Install a folder on the desktop containing five tests, one for each day of the week, Monday through Friday. The test corresponding to the day of the week is run at the beginning of the day to ensure that the system is working properly.
- Each test is the same, with the exception of checking the boot drive "C." Each day checks different blocks within drive "C."
- When the diagnostic check is activated, the system will exit to DOS, run the diagnostic and then return to Windows upon completion.
- Each test lasts approximately 15 min and writes a diagnostic report "AMIDIAG.log" to the AMIDiag directory on the "C" drive.
- This log should be printed out immediately and filed. It is a text file and can be opened in Notepad, Wordpad, or Microsoft Word.

Note: The AMIDIAG.LOG file will be overwritten when the next day's test is run. Print it out as soon as the diagnostic is run.

SAMPLE DIAGNOSTIC REPORT

```
AMIDiag Test report

NO DEVICE for SCSI Disk Read Test
NO DEVICE for SCSI Disk Write Test


=============================================================
BATCH STARTED 10/02/97 08:56:10
=============================================================


_____

***** PASS 1 CYCLE 1 *****
_____

NO DEVICE for SCSI Disk Read Test
NO DEVICE for SCSI Disk Write Test

[System: Basic Functionality Test]
TEST STARTED 10/02/97 08:56:12
32 Bit Register Read/Write: OK
32 Bit Stack Instructions: OK
32 Bit Data Access through FS,GS: OK
```

Bit Instructions: OK
32 Bit Extended FLAG Instructions: OK
Protected Mode Entry Instructions: OK
32 Bit Multiplication: OK
TEST PASSED
TEST ENDED 10/02/97 08:56:25

[System: CPU Protected Mode Test]
TEST STARTED 10/02/97 08:56:26
A20 LINE: OK
Protected Mode Instructions: OK
TEST PASSED
TEST ENDED 10/02/97 08:56:26

[System: Processor Speed Test]
TEST STARTED 10/02/97 08:56:27
CPU Speed is 200 MHz.
TEST PASSED
TEST ENDED 10/02/97 08:56:28

[System: Coprocessor Test]
TEST STARTED 10/02/97 08:56:29
Control Word Read/Write: OK
FSTENV Instruction Test: OK
FPU Stack Read/Write: OK
Integer Rounding Test: OK
Integer Loading Test: OK
FPU Tag Condition: OK
TEST PASSED
TEST ENDED 10/02/97 08:56:29

[System: DMA Controller Test]
TEST STARTED 10/02/97 08:56:30
DMA Controller 1 Registers: OK
DMA Controller 2 Registers: OK
DMA Controller Page Registers: OK
TEST PASSED
TEST ENDED 10/02/97 08:56:30

[System: Interrupt Controller Test]
TEST STARTED 10/02/97 08:56:31
Enable all Mask bits: OK
Disable all Mask bits: OK
Test for stray interrupts: OK
TEST PASSED
TEST ENDED 10/02/97 08:56:32

```
[System: Timer Test]
TEST STARTED 10/02/97 08:56:33
Set RTC and Timer Clock: OK
Checking Periodic Interrupts...
Calibrate Timer: OK
TEST PASSED
TEST ENDED 10/02/97 08:56:34

[System: Real Time Clock Test]
TEST STARTED 10/02/97 08:56:35
Set RTC and Timer Clock: OK
Checking Periodic Interrupts...
Calibrate Timer: OK
RTC Pattern Test: OK
TEST PASSED
TEST ENDED 10/02/97 08:56:36

[System: CMOS Validity Test]
TEST STARTED 10/02/97 08:56:37
Power Condition: OK
Checksum Condition: OK
Configuration Status: OK
Memory Size: OK
CMOS Time Status: OK
Divisor Bits: OK
Rate Selection Bits: OK
Update Cycle: OK
Calculated Checksum: OK
Shutdown Register RD/WR: OK
TEST PASSED
TEST ENDED 10/02/97 08:56:39

[System: PCI System Test]
TEST STARTED 10/02/97 08:56:40
PCI Bus Detected
PCI Bus Scanned: OK
PCI Device Access: OK
PCI Config. Verification: OK
TEST PASSED
TEST ENDED 10/02/97 08:56:40

[Memory: Random Memory Test]
TEST STARTED 10/02/97 08:56:41
TEST PASSED
TEST ENDED 10/02/97 09:09:27


======================================================================
BATCH ENDED 10/02/97 09:09:27

======================================================================
```

CHAPTER 6

IMAGE PROCESSING

What do we mean by image processing? This really starts after the pictures are taken and stored on the flash card. Now we can start working on it, in other words editing, enhancing, and analyzing the image.

Adobe Photoshop is not the only image editing/enhancement program on the market, but it is one of the most powerful programs available off-the-shelf. Off-the-shelf is important, because it is available to anyone. This is not a program that is proprietary to law enforcement or forensics, but one that was designed for graphics work and desktop publishing that just happens to work quite well for forensic images. It is a program that is widely accepted throughout the imaging world and has already been tested in a few court cases. Adobe has also agreed to testify about the program and its operation if needed.

All these points make it less likely that the application program itself would be questioned in court. This does not mean that what you do to an image in Adobe Photoshop won't be questioned. Adobe is a powerful program and there is a fine line between enhancement and manipulation of an image. Ensure that you know where the line is and don't cross over it.

ADOBE PHOTOSHOP

Adobe Photoshop 6.0 software has introduced the next generation of image editing with powerful new features that offer something for every user. One of the great additions for us in forensics is the enhanced layer control. New layer design features allow you to apply editable gradients, patterns, and solid colors, as well as color adjustments, to other layers.

Automation is another great feature, record your editing steps as an action, and then automatically playing it back on any file or batch of files. The expanded support now includes setting tool options and better batching for opening and renaming files. You can save actions as droplets, which trigger batch operations when you drag folders of images onto them.

These are just a few of the features that have been upgraded with the new version. We will discuss each of the features as we progress through the program. Adobe Photoshop is used around the world to create the highest quality images for many different uses. Most digital imaging experts will tell you that Adobe Photoshop is the center of the digital imaging universe. Its capabilities are really only limited by the technicians knowledge of the program.

CALIBRATING THE MONITOR

It would be a perfect world if the colors you see on the monitor were exactly what would be produced when printed on the color printer, but many times you will find this is not the case.

There are a number of reasons for this, one being, what you see on the monitor is RGB (red, green, and blue), considered the primary (additive) colors. These RGB colors must be converted to CMYK (cyan, magenta, yellow, and black), considered the secondary (subtractive) colors for printing. This requires calibrating the monitor to the printer and is another problem that you will encounter. We will discuss this more in detail later, but the bottom line is that all RGB colors cannot be reproduced on a CMYK printer.

The place to start is with what you see when you're working on an image, and that begins with the monitor. A properly calibrated monitor can help ensure that what you are seeing *more closely* resembles the printed output. Therefore, the first step recommended by Adobe is calibrating your monitor. Adobe has supplied as part of the Photoshop program an application called Adobe Gamma. This program is used to calibrate the monitor so that you are seeing colors as true as possible on your monitor under the lighting conditions you are working. The Adobe Gamma utility guides you with a wizard to help you calibrate the contrast and brightness, gamma (midtones), color balance, and white point of the monitor. This helps eliminate any color cast in your monitor display, sets your monitor grays as neutral as possible, and standardizes your display images on different monitors (whatever the combinations of monitor and video card). The settings are saved as an ICC profile for your monitor. You only need to set calibration and save it as an ICC profile once on your system, for all applications, unless you change any of the factors affecting calibration.

Ensure that the monitor has been turned on for at least 30 min and set the room lighting at the level you plan to maintain. Room lighting is very important; it is even helpful to install a makeshift hood around the monitor to keep out any extraneous light.

Turn off any desktop patterns and change the background color on your monitor to a light gray. This prevents the background color from interfering with your color perception. Before starting the calibration process, familiarize yourself with the controls for your monitor, brightness, contrast, etc.

The Adobe Gamma utility appears automatically if you are installing

Photoshop. If you cancel it during installation, it can be started later from the "Control Panel." Start the Adobe Gamma utility. The Adobe Gamma dialog box will appear as shown in Figure 6.1. Select the step-by-step wizard and click on the next button. The step-by-step option guides you through the entire process. A default profile will be loaded to describe your monitor if you haven't previously loaded your own profile as shown in Figure 6.2. The "load" button allows you to load a new profile. Click the "next" button to start calibrating the monitor and creating a new profile.

The first step is to adjust the brightness and contrast of your monitor. Set the contrast control to its highest setting using the guide in Figure 6.3. Adjust the brightness control to make the center box as dark as possible, but not black – keep the frame a bright white. Click the "next" button.

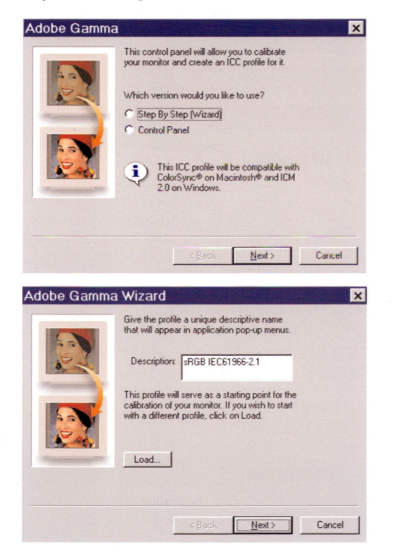

Figure 6.1

Figure 6.2

Figure 6.3

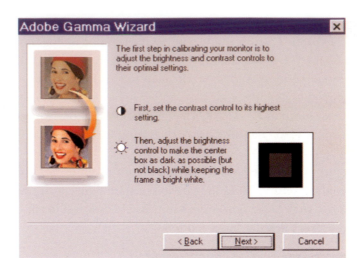

Figure 6.4

Figure 6.5

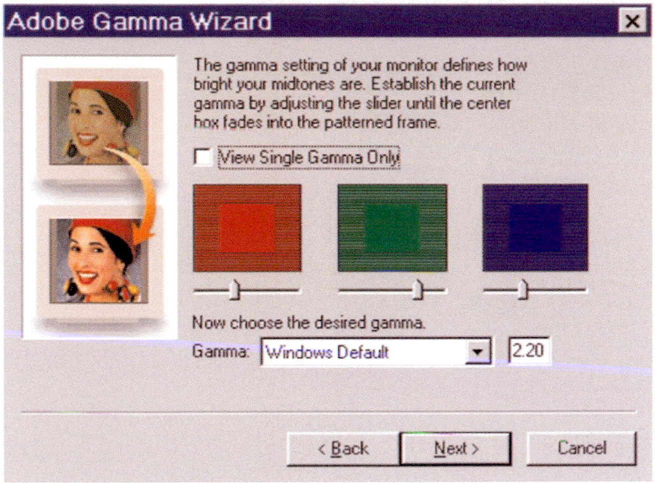

Figure 6.6

Click on the drop-down arrow shown in Figure 6.4 and select your monitor if it is listed. If it is not listed, choose Custom and click on the next button. The "Adobe Gamma" dialog box will open. There are two methods of adjusting gamma – we will explore each one.

Place a check in the box in Figure 6.5 next to "View Single Gamma Only" setting. This will allow you to adjust the gamma based on a single combined grayscale reading. Move the slider until the center box fades into the patterned frame.

In Figure 6.6 remove the check mark from the box next to "View Single Gamma Only." This will allow you to adjust the gamma based on the red, green, and blue readings. Adjust the slider under each box, until the center box matches the patterned frame.

The gamma setting of your monitor defines how bright the midtones are. Choose the target gamma you want – the default target gamma in Windows is 2.2. Click "next" to proceed.

In the dialog box that opens (Figure 6.7) you can set the white point of the monitor – leaving it at its default "5000° K" is usually close enough for most monitors. The measure button allows you to measure and set the white point for your monitor. If you decide to measure the white point – all ambient light must be turned off then follow the instructions. Click the "next" button.

The box in Figure 6.8 allows you to set the "Adjusted White Point." This normally will be left at its default (Same as Hardware). The only time you would change this is if you had a controlled light source for viewing your prints – then set it to match the light source. Click the "next" button to proceed.

In Figure 6.9 you can check the before and after effects of your changes by clicking on the radio buttons. Click the "Finish" button.

Figure 6.7

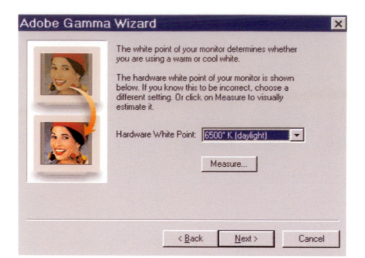

Figure 6.8

Figure 6.9

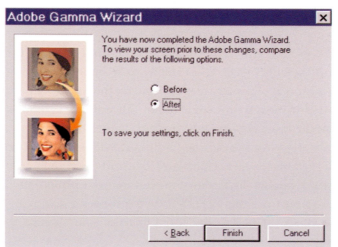

Click the "Save" button on the final dialog box and the profile will be saved. Use this as the ICC profile for the monitor, unless conditions change where the monitor is located.

Note: if anyone changes any of the settings on the monitor, such as brightness or contrast, the monitor calibration will be voided.

PHOTOSHOP MENUS

Figures 6.10 to 6.19 show the main menus for working with Photoshop.

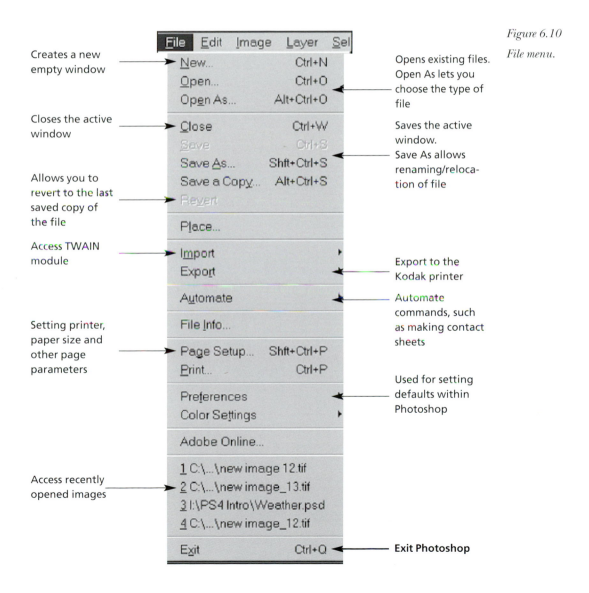

Creates a new empty window

Closes the active window

Allows you to revert to the last saved copy of the file

Access TWAIN module

Setting printer, paper size and other page parameters

Access recently opened images

Opens existing files. Open As lets you choose the type of file

Saves the active window. Save As allows renaming/relocation of file

Export to the Kodak printer

Automate commands, such as making contact sheets

Used for setting defaults within Photoshop

Exit Photoshop

Figure 6.10
File menu.

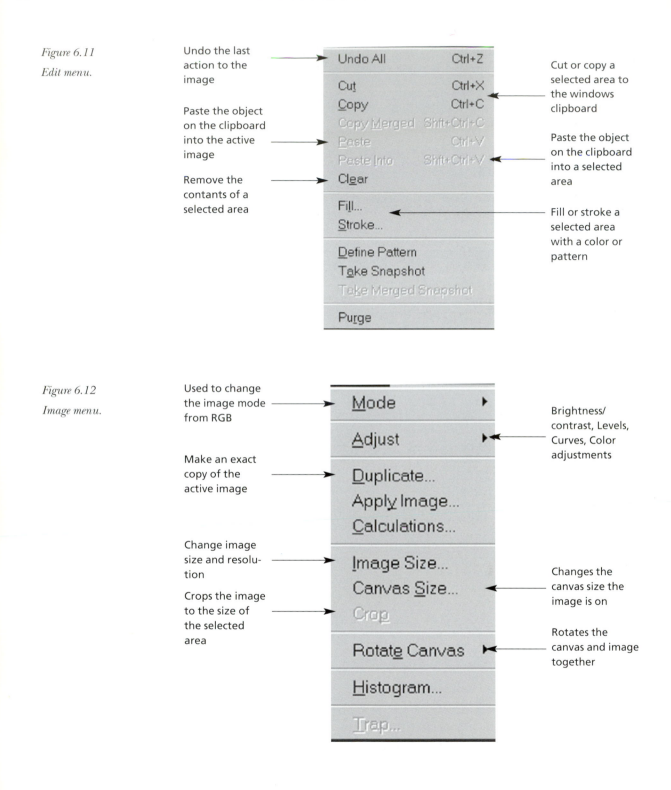

Figure 6.11
Edit menu.

Undo the last action to the image

Paste the object on the clipboard into the active image

Remove the contants of a selected area

Cut or copy a selected area to the windows clipboard

Paste the object on the clipboard into a selected area

Fill or stroke a selected area with a color or pattern

Undo All   Ctrl+Z
Cut   Ctrl+X
Copy   Ctrl+C
Copy Merged   Shift+Ctrl+C
Paste   Ctrl+V
Paste Into   Shift+Ctrl+V
Clear

Fill...
Stroke...

Define Pattern
Take Snapshot
Take Merged Snapshot

Purge

Figure 6.12
Image menu.

Used to change the image mode from RGB

Make an exact copy of the active image

Change image size and resolution

Crops the image to the size of the selected area

Brightness/ contrast, Levels, Curves, Color adjustments

Changes the canvas size the image is on

Rotates the canvas and image together

Mode

Adjust

Duplicate...
Apply Image...
Calculations...

Image Size...
Canvas Size...
Crop

Rotate Canvas

Histogram...

Trap...

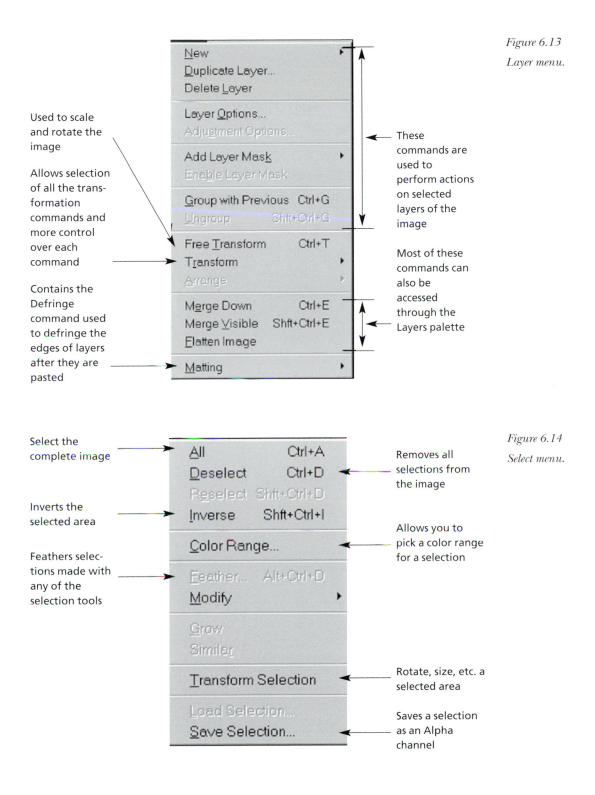

Used to scale and rotate the image

Allows selection of all the transformation commands and more control over each command

Contains the Defringe command used to defringe the edges of layers after they are pasted

These commands are used to perform actions on selected layers of the image

Most of these commands can also be accessed through the Layers palette

Figure 6.13
Layer menu.

New	▶
Duplicate Layer...	
Delete Layer	
Layer Options...	
Adjustment Options...	
Add Layer Mask	▶
Enable Layer Mask	
Group with Previous	Ctrl+G
Ungroup	Shft+Ctrl+G
Free Transform	Ctrl+T
Transform	▶
Arrange	▶
Merge Down	Ctrl+E
Merge Visible	Shft+Ctrl+E
Flatten Image	
Matting	▶

Select the complete image

Inverts the selected area

Feathers selections made with any of the selection tools

Removes all selections from the image

Allows you to pick a color range for a selection

Rotate, size, etc. a selected area

Saves a selection as an Alpha channel

Figure 6.14
Select menu.

All	Ctrl+A
Deselect	Ctrl+D
Reselect	Shft+Ctrl+D
Inverse	Shft+Ctrl+I
Color Range...	
Feather...	Alt+Ctrl+D
Modify	▶
Grow	
Similar	
Transform Selection	
Load Selection...	
Save Selection...	

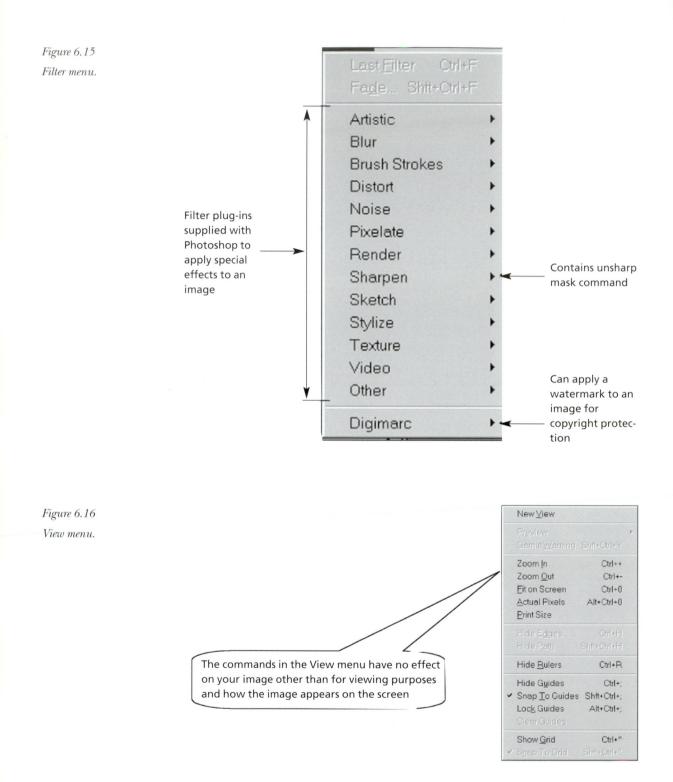

Figure 6.15
Filter menu.

Filter plug-ins supplied with Photoshop to apply special effects to an image

Contains unsharp mask command

Can apply a watermark to an image for copyright protection

Figure 6.16
View menu.

The commands in the View menu have no effect on your image other than for viewing purposes and how the image appears on the screen

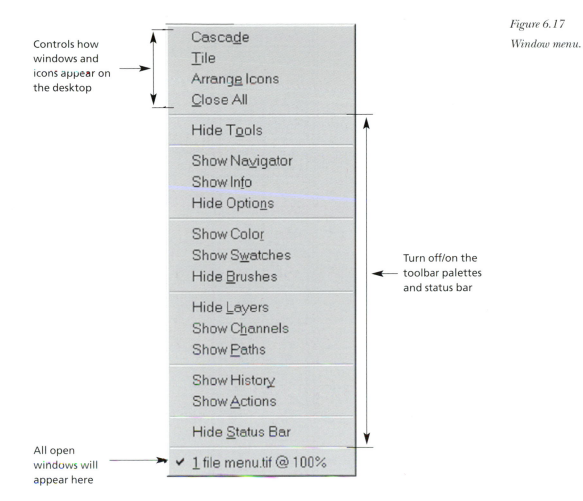

Controls how windows and icons appear on the desktop

All open windows will appear here

Turn off/on the toolbar palettes and status bar

Figure 6.17
Window menu.

Figure 6.18
Photoshop toolbar.

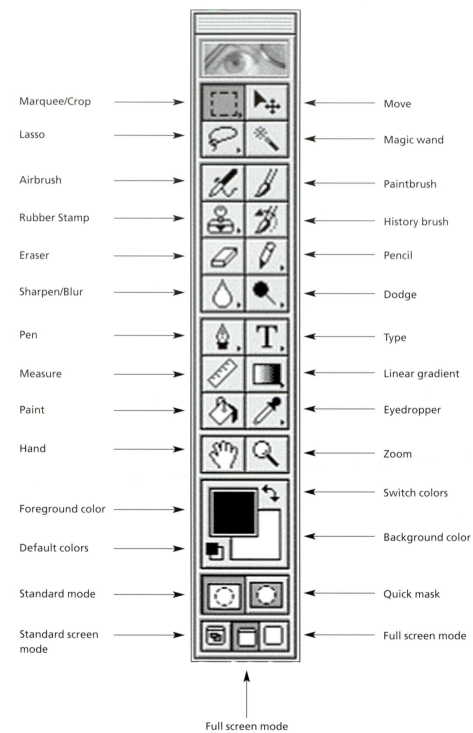

Marquee/Crop

Lasso

Airbrush

Rubber Stamp

Eraser

Sharpen/Blur

Pen

Measure

Paint

Hand

Foreground color

Default colors

Standard mode

Standard screen mode

Move

Magic wand

Paintbrush

History brush

Pencil

Dodge

Type

Linear gradient

Eyedropper

Zoom

Switch colors

Background color

Quick mask

Full screen mode

Full screen mode with menu bar

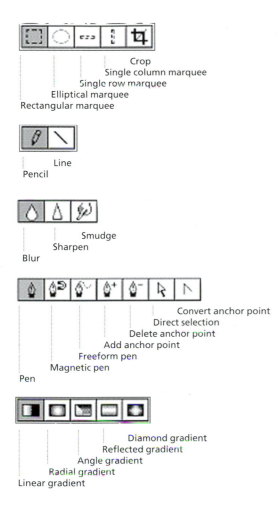

Crop
Single column marquee
Single row marquee
Elliptical marquee
Rectangular marquee

Line
Pencil

Smudge
Sharpen
Blur

Convert anchor point
Direct selection
Delete anchor point
Add anchor point
Freeform pen
Magnetic pen
Pen

Diamond gradient
Reflected gradient
Angle gradient
Radial gradient
Linear gradient

Figure 6.19
Toolbar flyouts.

Magnetic lasso
Polygon lasso
Lasso

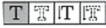

Pattern stamp
Rubber stamp

Sponge
Burn
Dodge

Vertical type mask
Vertical type
Type mask
Type

Color sampler
Eyedropper

ADOBE PHOTOSHOP DESKTOP

Note: You must study and become familiar with the toolbox and palettes before working on an image.

START ADOBE PHOTOSHOP

Open IMAGE_1 from your CD and let's explore the controls we have on the desktop before beginning any enhancements on an image. If the rulers do not appear around the image, go to the View > Show Rulers command.

At default Photoshop places the toolbox on the left of the desktop and the palettes on the right as shown in Figure 6.20. The toolbox contains all the tools necessary to work on an image and the palettes help you monitor and modify images. You can display or hide them as you work. By default, they appear in

Figure 6.20

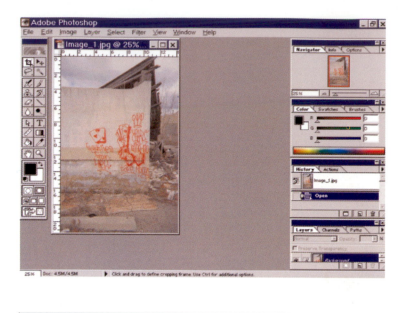

Figure 6.21

stacked groups. To show or hide a palette choose the appropriate Window > Show or Window > Hide command.

Both the toolbox and palettes are floating. Use the mouse, click in the title bar of the toolbox and drag it anywhere on the desktop. The palettes can be moved the same way, allowing you to set up the desktop any way you want to.

The palettes can also be modified to fit your particular use. Click on the tab of the history palette and drag it out by itself. Now click on the channels tab on top of the history palette. You now have created a palette with the history and channels together. Don't worry about getting them out of their default position, Photoshop allows you to reset them quickly using File > Preferences > General > Reset Palette Locations To Default. The other part of the Photoshop desktop that many of us ignore and fail to use is the status bar at the bottom. There is a multitude of information available right in the status bar (Figure 6.21). At its default it is broken into four sections. On the left it shows the zoom factor of the open image; this is also shown in the title bar of the image. This will vary depending on the resolution of your monitor. This has nothing to do with print size, just viewing. Next it shows two numbers with a slash between them; the document (image) file size is on the left and the document (image) file size if it were saved with any channels or layers is on the right. They will be the same size on a new image, but we will point out how the right number increases as we work on an image. To the right of the document size is an arrow; clicking on it will produce a fly-up menu as shown in Figure 6.22. This allows you to change the information that appears where the document sizes are shown.

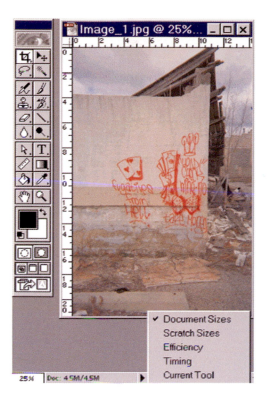

Figure 6.22

Click on scratch sizes in the fly-up menu. The left number shows the memory (RAM) consumed by all open images. The right number shows the amount of memory (RAM) available to the program. (*Note:* When the left number exceeds the right number, there is no memory left, and Photoshop must employ virtual memory that it has set aside on your hard drive. This will slow down your system tremendously.)

Three other commands also appear in the fly-up menu:

- Efficiency: if anything other than 100% appears, Photoshop is using virtual memory and there will be a drop in performance.
- Timing: Photoshop displays the total time in seconds it took to accomplish the last task.
- Current Tool: the current tool is displayed in the status bar.

After you have explored the different setting, return the setting to Document Sizes. If you look to the right of the arrow in the status bar, you will see information pertaining to the tool you have selected in the toolbox. It shows hot-keys from the keyboard that can be used in conjunction with the tool when working on an image. This information will change as you select different tools.

Since 100% size on the monitor does not reflect the printed size of the

Figure 6.23

Figure 6.24

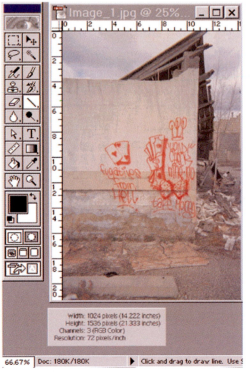

image, you must use the "Document Size" box or the Print Size command, to see how an image will print. Position the pointer over the document size box in the status bar and hold down the left mouse button. The physical size of the image is shown, as it would print on an 8.5 × 11 inch page (Figure 6.23). Notice that it indicates it will take the entire page, but look at the rulers; your image is very large. This will change in appearance, depending on what printer you have set up in the printer setup, but it will always show physically how the image will print.

Physical print size can also be seen by using the View > Print Size command. When the Print Size command is invoked, the image resized the percentage to show the actual size.

(*Example:* If you have a monitor with a resolution of 72 ppi and an image resolution of 120 ppi, it would resize to 60%. On a monitor with higher resolution than 72 ppi, the image would appear smaller than the printed size.)

Now hold down the Alt key while holding down the mouse button on the document size box. The dialog box appears in Figure 6.24. This will show the image size in inches and pixels as well as the type and resolution of the image. Nothing can be adjusted in this box, it is only for viewing.

BASIC TOOLS FOR ENHANCEMENT

As we know, Adobe Photoshop has a tool for just about anything you would want to adjust in an image, but the truth is, most of the time it is not necessary to do any fancy adjustments. As a rule, most of the time you're going to find out that your images just need basic adjustments, the same as we've been doing in the darkroom for years on conventional film images. The difference lies in the order we do them in; this is important when working with digital images and should be adhered to for each image. Step-by-step procedures are listed below:

1. Sizing the image
2. Cropping the image
3. Setting the contrast/brightness/gamma
4. Adjusting the color balance
5. Sharpening the image
6. Printing or archiving the image

We will take each of these procedures and learn the tools to use and the do's and don'ts that apply to each.

Figure 6.25

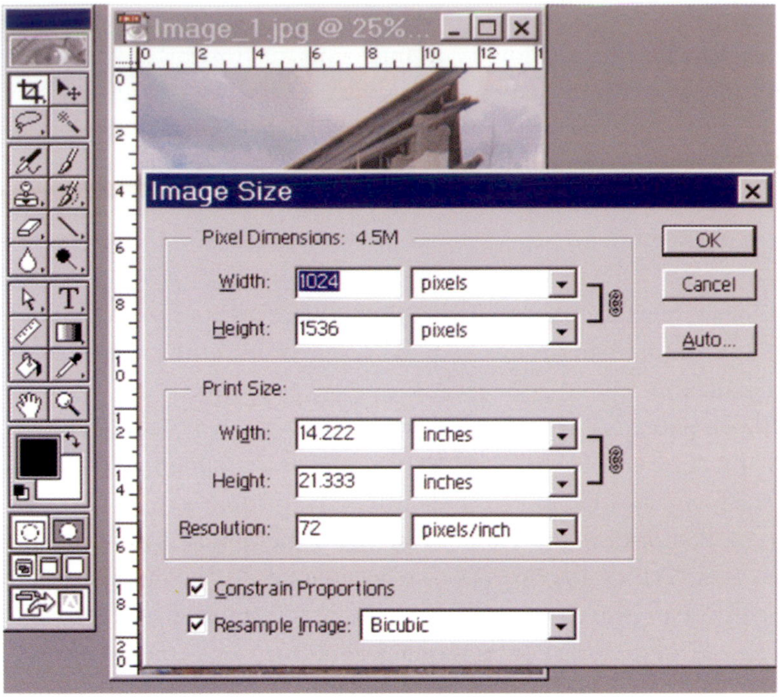

SIZING THE IMAGE

Choose the Image menu > Image Size command and the image size dialog box will open (Figure 6.25). We will explore the image size dialog box extensively, because it can be the most dangerous control in Photoshop if not used correctly. There are two sections to this dialog box, pixel dimensions on top and print size on the bottom. They each have a specific purpose.

As a rule-of-thumb, pixel dimension controls are used for resizing an image for viewing on a monitor, print size controls are used for sizing your finished print.

Click on the box next to Resample Image a couple of times; notice the pixel dimension section is grayed out when the resample is off. Total pixels in the image will remain the same if you resize the image. Resampling refers to changing the pixel dimensions (and therefore file size) of an image. When you down-sample (or decrease the number of pixels), information is deleted from the image. When you resample up (or increase the number of pixels), new pixel information is added based on color values of existing pixels. You specify an interpolation method to determine how pixels are added or deleted.

In Figure 6.26 at the bottom of the dialog box there is a drop-down arrow allowing you to set the method of interpolation. There are three choices:

1. Nearest Neighbor for the fastest, but least precise, method. This method can result in jagged effects, which become apparent when distorting or scaling an image or performing multiple manipulations on a selection.
2. Bilinear for a medium-quality method.
3. Bicubic for the slowest, but most precise, method, resulting in the smoothest tonal gradations.

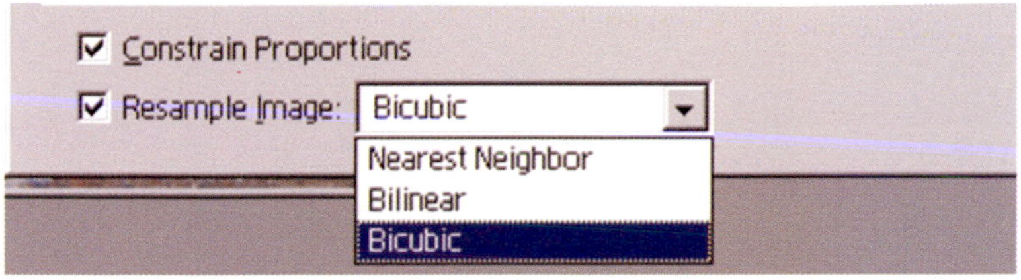

Figure 6.26

Bicubic is by far the best method and should be used on all continuous tone images, but there are limits to how much you can enlarge a digital image and maintain quality. *Beware* – Photoshop does a good job of doubling the size of an image, but for anything above doubling, you will start to see a color shift in the image and also loss of sharpness. Setting the image size should be the first step in working on any digital image. This ensures that the pixels you enhance in later steps are the pixels that will be in the final image.

Notice in the image size dialog box (Figure 6.27) how large the image is and the low resolution. This is typical of an image from a point-and-shoot digital camera. Assuming we are going to print the image on a high-resolution printer

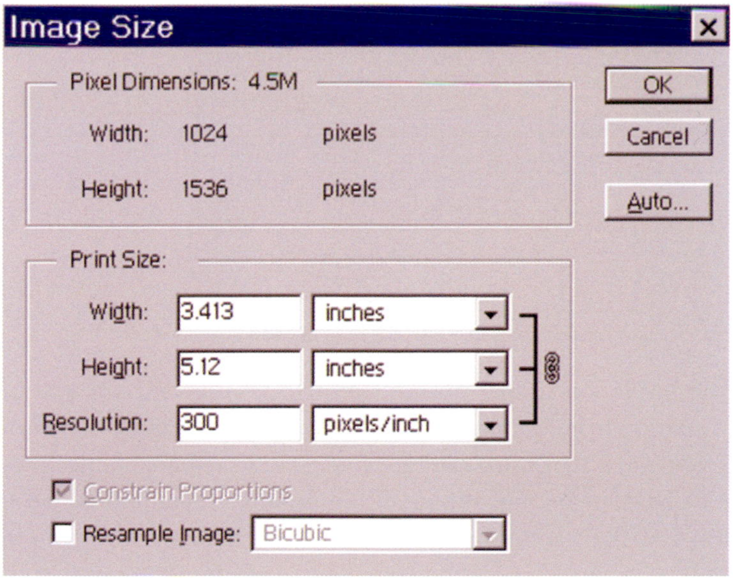

Figure 6.27

Figure 6.28

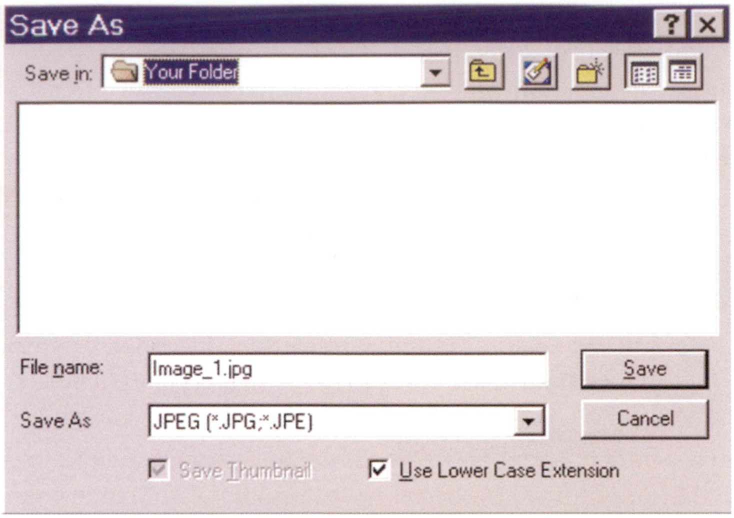

Figure 6.29

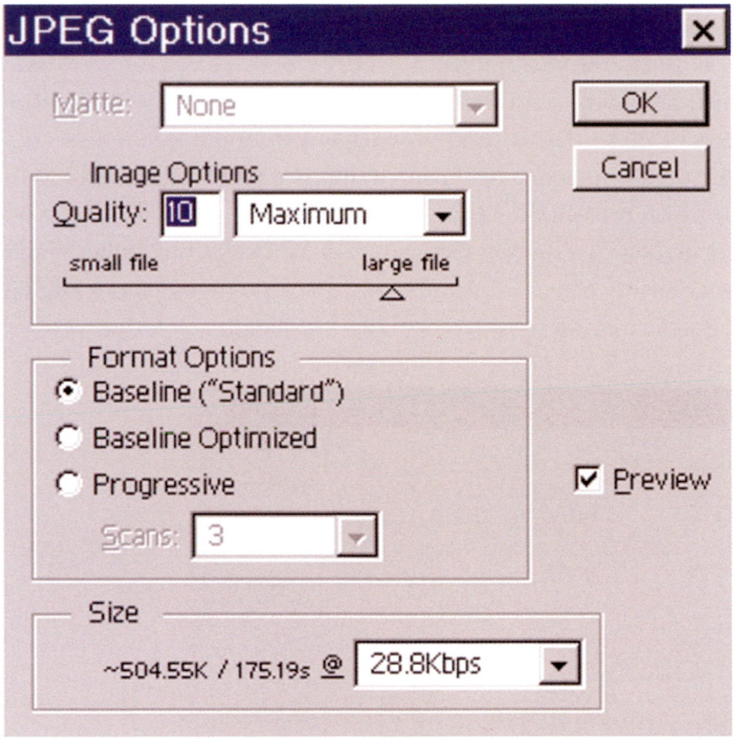

we need to set the resolution to 300 ppi. In order to do this and not add or subtract any information from the image we must turn the resample off. Turn the resample off and enter 300 in the resolution box (Figure 6.27). Notice how the image dimensions changed, but the pixel dimensions remained constant. The file size is still 4.5 MB; therefore we have not changed any information in the image. Click OK in the dialog box, the image is resized. It did not physically change size on the desktop, because the image has the same amount of pixels, but if you observe the rulers you will see the new image size.

It's time to save the image. You can't save it on the CD, so build yourself a folder somewhere on your hard drive and use it to save all your images in as you work. Choose the File > Save As command (Figure 6.28). Find "Your Folder," do not change the name, and save it as a JPEG. Click on the Save button. When the JPEG Options box opens (Figure 6.29) set the quality to 10 and click OK.

Now let's see how to enlarge an image; remember doubling the size is about the limit. Choose the Image > Image Size command again to open the dialog box. To enlarge an image you must turn on the resample; notice Pixel Dimensions is now highlighted, which will allow them to be changed (Figure 6.30).

The height is 5.12 inches, so we can enlarge it to 10 inches and still be less than double. Enter 10 in the height box. The width will change automatically, because the constrain proportions is turned on. Notice the pixel dimensions; the file size went from 4.5 MB to 17.2 MB and the total pixels almost doubled. Where are these extra pixels coming from? Photoshop is going to create them by resampling and interpolation.

Click OK to resize the image. The image now is much larger on your monitor. Depending on how large a screen your monitor has, you may not be able to see all of it without scrolling. The reason for this is that you increased the pixels in your image. Let's save this image to your folder. Rename it IMAGE_1E, so we will know it is the enlarged version. We're going to print both these images later to see if we can see any difference.

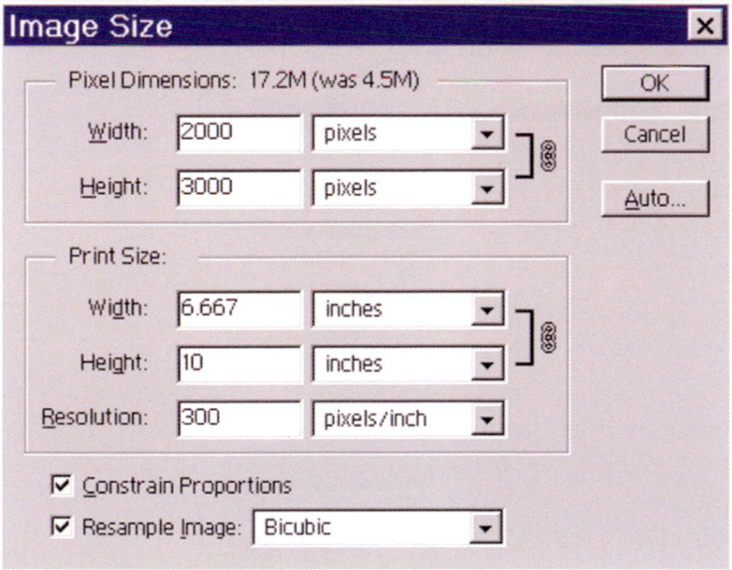

Figure 6.30

Zoom tool

Now is a good time to discuss the zoom tool. Select the zoom tool from the toolbox. Move the cursor onto the image; a plus sign (+) appears in the cursor. Click one time in the image. Watch the zoom factor in the title bar of the image, and continue clicking until you have reached 800% zoom (Figure 6.31). At this zoom factor you should easily be able to see the pixels in the image.

Figure 6.31

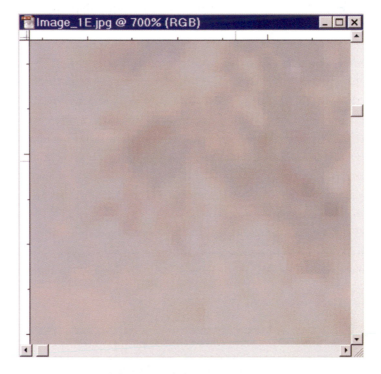

Figure 6.32

Click on the zoom options tab in the palettes (Figure 6.32). Place a check mark in the Resize Windows To Fit Box. Each time you zoom the picture box will resize with the image.

Place the cursor inside the image and hold down the Alt key; a minus sign (−) appears in the cursor. Click in the image, it zooms out. Continue clicking until you can see the complete image.

For those of you that like to use the keyboard, holding the Ctrl key and pressing the "+" or "−" key will also zoom in or out on the image. Experiment with it a couple of times. When you are done, return to where you can see the complete image. Photoshop also has a navigator palette that is good for zooming and rapidly moving around the image. Move the palette with the navigator over next to the image (Figure 6.33). Click on the navigator tab. Moving the slider at the bottom of the palette or clicking the mountain icons will change the zoom factor. Zoom in on the image, using the slider. Notice the red box in the image. Place the cursor in the red box and a hand will appear. Hold the mouse button down and scroll through the image.

Return the navigator palette to its original location. Click on the "×" in the upper right corner of the image window to close the image.

Open IMAGE_1 from your folder.

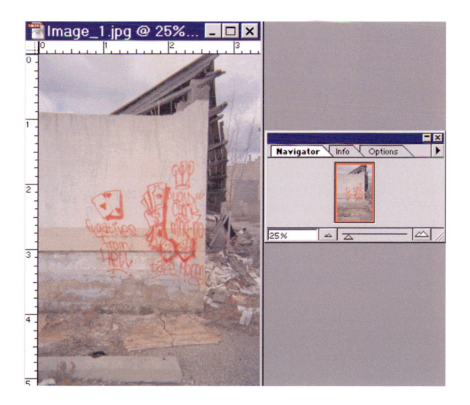

Figure 6.33

CROPPING AN IMAGE

To put it bluntly, you shouldn't crop a digital image at all in any software program. Think about it for a minute; when you crop, valuable pixels are being eliminated – pixels you can't afford to lose. All cropping when using a digital camera should be done *in the camera* or if you're scanning, *crop in the scanner*. As photographers, we have become lazy; film is so good we don't have to worry about filling the frame when we shoot – just crop and enlarge it in the darkroom.

We do not have this luxury with digital images. We already discussed sizing an image and know the problems we have with interpolation and enlargement of a digital image. Until digital cameras reach the resolution of film, we've got to train ourselves to shoot smarter and use every pixel available to our best advantage. But, with all that said, we know we're going to do some cropping in Photoshop from time-to-time, so let's see how to use the tools and get the best results.

Photoshop provides two ways to crop an image:

1. The Image > Crop command discards the area outside of a rectangular selection and keeps the same resolution as the original. This discards pixels and reduces the physical size of the image.
2. The crop tool lets you crop an image by dragging over the area you want to keep. The advantage of using the crop tool is that you can rotate and resample the area as you crop.

Remember what we said about resampling earlier. Let's assume that you want to crop everything out of the picture but the graffiti. Select the rectangular marquee tool from the toolbox. Move onto the image, upper left of the graffiti, hold down the mouse button, and drag down diagonally until the marquee encompasses the graffiti as in Figure 6.34. Release the mouse button.

Choose the Image Menu > Crop command. Everything outside the marquee is removed. Note the picture size; it is reduced and the file size is smaller, but the resolution remains the same at 300 ppi. Check it by holding the Alt key and clicking "Doc: size" in the status bar. The image is the same quality (330 ppi) but is smaller.

Revert command

Now is a good time to check out the Revert command. An image can easily be restored to the last-saved version by using the Revert command. This is just one of the many ways we will discuss to return to a previous state in our image, very good for correcting mistakes. Choose the File > Revert command. The image reverts back to its state prior to cropping.

Figure 6.34

Crop tool

Although the marquee is a rapid method of cropping, you have very little control over the tool. The crop tool is the preferred method of cropping. The crop tool is hidden behind the marquee tools; hold down the mouse button, drag to select the crop tool. After it is selected, double-click it to open the crop options palette. First of all, let's look at the crop options palette in Figure 6.35. There are two methods of cropping using the crop tool: freehand and cropping to size.

Freehand cropping with the crop tool is the same as using the Image > Crop command, except that you can resize your crop and rotate it. If necessary remove the check mark from the Fixed Target Size option in the palette. Move the cursor onto the image and crop the graffiti as you did with the marquee tool.

Notice the difference, now you have eight handles around the marquee for adjustments (Figure 6.36).

The handles are used for resizing the crop area. If it needs adjustments, place the cursor over any handle; it will become a double-headed arrow. Drag the handle to change the cropped area. To move the marquee, place the cursor inside the marquee. Do not place it on the bull's-eye in the center. The cursor becomes a pointer. Hold down the mouse button and drag to move the

Figure 6.35

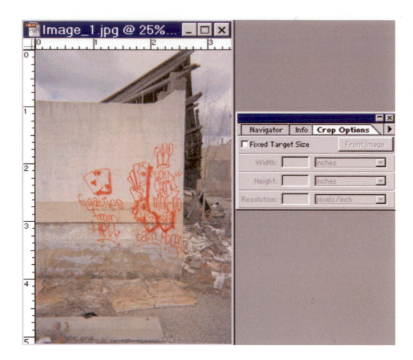

Figure 6.36

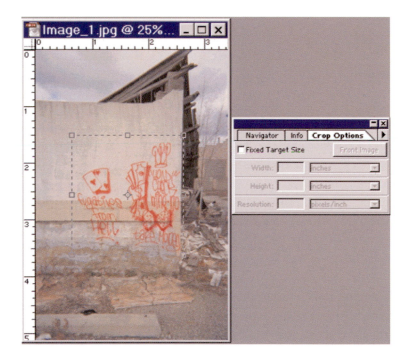

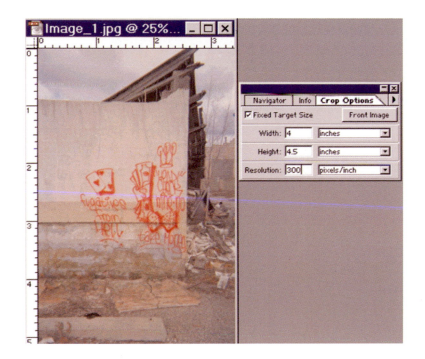

Figure 6.37

marquee. To rotate the crop area position the cursor outside the marquee. The cursor becomes curved with arrows on both ends. Hold down the mouse button and drag to rotate the marquee. Notice that it rotates around the bull's-eye in the center. By placing the mouse on the bull's-eye and moving it you can change the pivot point of the rotation. Pressing the Esc key cancels the crop, pressing the Enter key completes the crop. Press the Esc key to remove the crop marquee.

Now let's crop the image to a particular size. Place a check mark in the fixed target size box (Figure 6.37). Click the Front Image button to see the size and resolution values of the current image. Choose inches for the unit of measurement from the menus. Enter a value of 4 in the width box and a value of 4.5 in the height box. Leave the resolution at 300 ppi.

Move the cursor onto the image and place a marquee around the graffiti. Notice that the proportion is constrained as you drag. Release the mouse button. Note that there are only four handles for adjustments, because the proportion is constrained (Figure 6.38). Adjusting the crop is the same as it was with the freehand crop.

The Fixed Target Size option constrains the file size of the image. If you specify size, but not resolution, the resolution changes automatically to compensate for the size change. If you specify resolution but not size, the size changes automatically to compensate for the resolution change.

Beware: there is no Resample on/off control in this dialog box. Photoshop

Figure 6.38

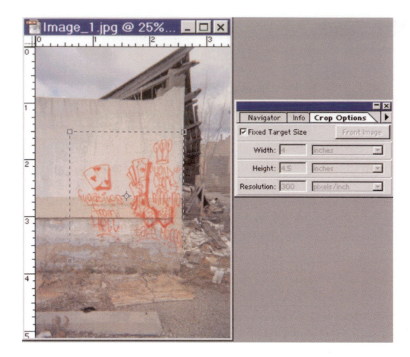

assumes you know what you're doing and resamples (interpolates) the image as necessary to meet the requirements that you entered in the boxes.

Press the Esc key to cancel the crop. Move the Crop Options palette back to its original location.

HISTOGRAM

Before we begin enhancing the tones in the image, let's explore the histogram to see what we are working with in the image. It's important to determine whether your image has sufficient detail to produce high-quality output. The higher the number of pixels in an area, the greater the detail. Bad photographs or bad scans can be difficult, if not impossible, to correct. Too many color corrections can also result in loss of pixel values and too little detail.

Choose the Image > Histogram command. A histogram of the active image will open (Figure 6.39). The histogram graphs the number of pixels at each brightness level in an image. The *x*-axis of the histogram represents the color values from darkest (0) at the far left to brightest (255) at the far right; the *y*-axis represents the total number of pixels with a given value. It can show you whether an image contains enough detail to make a good correction. It also gives a quick picture of the tonal range of the image, or the image key type. A low-key image has detail concentrated in the shadows; a high-key one has detail concentrated in the highlights. An image with full tonal range has a high number of pixels in all these areas. Identifying tonal range helps determine

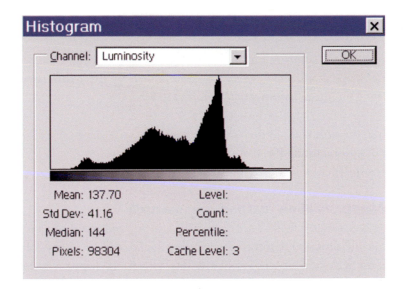

Figure 6.39

appropriate tonal corrections. The numerical values at the lower left of the Histogram dialog box display statistical information about the color values of the pixels:

- Mean is the average brightness value.
- Standard deviation (Std Dev) represents how widely the values vary.
- Median shows the middle value in the range of color values.
- Pixels represents the total number of pixels used to calculate the histogram.
- Cache Level shows the setting for the image cache. If the Use Cache for Histograms option is selected in the Memory & Image Cache Preferences dialog box, the histogram is based on a representative sampling of pixels in the image (based on the magnification), rather than on all of the pixels.

If you move the mouse over the histogram you can see how many pixels are at each level. The values at the lower right of the dialog box change to display the gray level (Level) of the point (from 0 to 255), the total number of pixels at that level (Count), and the percentage of pixels at or below that level (Percentile).

A quick look at the histogram in Figure 6.39 tells us that this image has no true blacks or whites, because there are no pixels at these levels. Study the histogram for a moment and get familiar with it. Click OK to close the histogram.

LEVELS ADJUSTMENTS

Now let's look at the levels adjustments and see how we can make the image better. Choose the Image > Adjust > Levels command. There are three different methods of adjusting an image within the Levels dialog box.

- Manually drag the sliders to set the adjustments.
- Use the Auto button and let Photoshop do the adjustments.
- Use the eyedroppers to sample the tones.

Setting the highlights, shadows, and midtones manually

The Levels dialog box (Figure 6.40) has two sections, input and output levels, and each has a specific function. Used correctly most images can be enhanced just as you would do in the darkroom. First of all let's look at the input section with the histogram. You will note that the histogram shows exactly what you saw in the previous histogram, the shadows (shown on the left), the midtones (shown in the middle), and the highlights (shown on the right). The difference in this dialog box is that you can make adjustments to the image. The levels sliders let you gradually adjust the brightness, contrast, and midtones in an image. By adjusting the midtones, you can change the brightness values of the middle range of gray tones without dramatically altering the shadows and highlights.

Figure 6.40

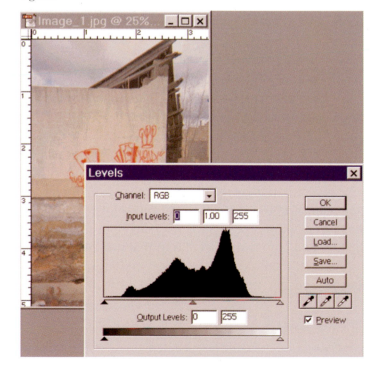

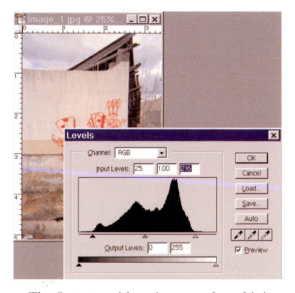

Figure 6.41

The first step with an image such as this is to spread the tones out by forcing a white and a black in the image. Move the (left) shadow levels input slider to the first group of pixels of the histogram. Repeat this procedure with the (right) highlight levels input slider as shown in Figure 6.41. Doing so maps these pixels – the darkest and lightest pixels in each channel – to black and white, increasing the tonal range of the image. The corresponding pixels in the other channels are adjusted proportionately to avoid altering the color balance.

The gamma control in the center of the input levels allows you to adjust the brightness of the midtones without affecting the shadows or highlight areas. Look at your image and adjust this to get the correct density.

You will find that the levels tool is the easiest to use and as a rule will perform all the adjustments you will need. But remember, study the histogram of the image; there must be pixels to adjust. It's just like film, if it is underexposed or overexposed excessively, there's very little you can do with it in the darkroom.

Now about those output sliders at the bottom of the box; these should be adjusted only after the input sliders are set, if necessary. The output levels sliders are used to reduce contrast in the highlights and shadows of the image. Sliding the left slider to the right reduces shadows. Sliding the right slider to the left reduces the highlights. Swapping the sliders will completely reverse the tones in your image (make it a negative). Click the Cancel button to close the dialog box without making any changes.

Using the Auto Levels button

Choose the Image Menu > Adjust > Levels command.

The Auto button in the Levels dialog box performs automatically the equivalent of the levels slider adjustment you just performed. It defines the lightest and darkest pixels in each channel as white and black and then redistributes the intermediate pixel values proportionately.

By default, the Auto feature clips the white and black pixels by 0.5%, that is, it ignores the first 0.5% of either extreme when identifying the lightest and darkest pixels in the image. This clipping of color values ensures that white and black values are based on representative rather than extreme pixel values.

Click on the Auto button; the image probably looks pretty close to what it did when you adjusted it manually.

The Auto feature gives good results when an image with an average distribution of pixel values needs a simple contrast adjustment. However, adjusting the levels controls manually is more precise. Click the Cancel button to close the dialog box without making any changes.

Using the eyedroppers

To use the eyedroppers to adjust the image you must first understand and be able to use the Info palette to read the levels in the image. Drag the Info palette out next to the image as shown in Figure 6.42. Choose the Image > Adjust > Levels command. Move the Levels dialog box off to the side; you need to see the complete image. Move the cursor out onto the image; it becomes an eyedropper. As you move it through the image, watch the RGB values in the Info palette. These are the only values we are interested in at this point, because it is an RGB image. A true white would have a reading of R255, B255, G255. A true black would have a reading of R0, B0, G0, but we already know there is no true white or black in the image, because of the histogram.

When using the eyedroppers it is important to identify a truly representative highlight, shadow and midtone (18% gray) area. Otherwise the tonal range may be expanded unnecessarily to include extreme pixel values that don't give the image detail. A highlight area must be a printable highlight, not specular

Figure 6.42

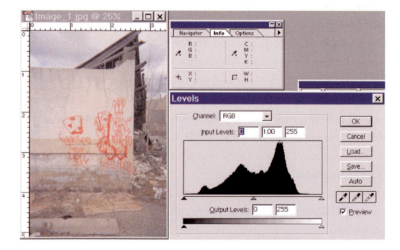

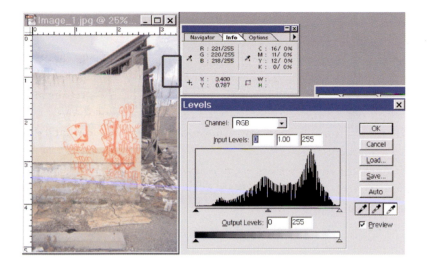

Figure 6.43

white. Specular white has no detail, and so no ink is printed on the paper. For example, a spot of glare is specular white, not a printable highlight. The bottom line is that the highlight you choose must have detail. In most situations when you are printing on white paper, you can achieve a good highlight in an average-key image using RGB readings of 244, 244, 244. As with white, when you're printing on white paper you can usually achieve a good shadow in an average-key image using RGB settings of about 10, 10, 10. This selects a dark portion of the image that still has some detail.

(*Note:* when using these eyedroppers, the (RGB) settings you select must be very close to equal. If not you will get a color shift in the image.)

Select the white eyedropper (on the right) (Figure 6.43). Move out onto the image and sample the highlights. The boxed area will probably be about as close as you will get to a white in this image. The RGB values should be in the 220 range. Click the eyedropper and watch the image; notice that it forced white pixels in the histogram.

Select the black eyedropper (on the left) (Figure 6.44). Move it into the area of the white box. The RGB values will be in the 30s. Click the eyedropper and watch the image and the histogram. The image now has white and black pixels.

Next is setting the midtones and eliminating color casts using the gray eyedropper. You can use the gray eyedropper button in the Levels dialog box to eliminate color casts in the midtones. But for precise control over midtone adjustments, use the levels midtone sliders. The ideal way to do this is to have an 18% gray card in your picture, but we know this isn't always possible. Like with the black and white eyedroppers, we must find a neutral tone in our picture. You can usually achieve a neutral midtone in an average-key image using RGB settings of about R127, B127, G127 or find a K value in CMYK of 18%. This selects a midtone portion of the image that has neutral tones.

Figure 6.44

Figure 6.45

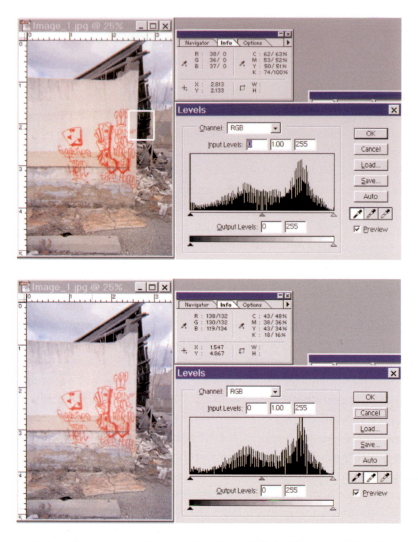

Select the gray eyedropper (in the middle) (Figure 6.45). Move the eyedropper over the parking block in the bottom of the picture. Click when you find 18% gray as shown in the K value of the info palette.

If used properly this will do a good job of color balancing an image. Midtone density still may have to be adjusted by using the slider.

This finishes the discussion of the levels tool. Hold the Alt key and click on the Reset button in the dialog box.

Using the sliders as we discussed earlier, adjust the image manually. When you have finished, click OK to apply the images.

Save and close the image.

BRIGHTNESS/CONTRAST COMMAND

Now let's look at a somewhat simpler tool called Brightness/Contrast Adjustment. I want to emphasize right from the beginning that this adjustment should never be used on a continuous tone image (photograph). Even Adobe will tell you this. The reason for this is that it does not allow the individual adjustments of the midtones of an image that are available in the levels tool.

It is the easiest way to make simple adjustments to the tonal range of the image. Unlike Levels, this command adjusts all pixel values in the image at once – highlights, shadows, and midtones, but remember it is *not recommended for high-end output*.

Now this is fine if there are no midtones, just black and white, or the midtones do not need adjustment, so think for a moment what kind of image this could be. The first one that comes to mind is fingerprints and possibly some documents. Just keep in mind, if the image needs midtone adjustments, do not use this tool.

Now let's see how to use it. Open a new image: choose the File > Open > Fingerprint.jpg from your CD (Figure 6.46). This is a poor Polaroid print that has been scanned. It is very hard to identify any munica points, because of the poor

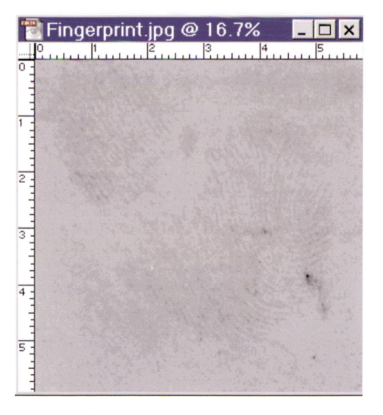

Figure 6.46

Figure 6.47

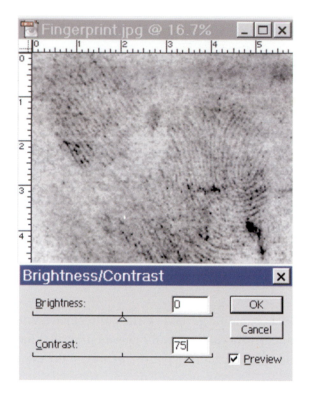

separation between the ridges and valleys. One of the ways to improve this separation is to increase the contrast of the image.

Choose the Image > Adjust > Brightness/Contrast command. The Brightness/Contrast dialog box will open. Moving the sliders adjusts the brightness and contrast. Dragging to the left decreases the level and to the right increases it. The number at the right of each slider value displays the brightness or contrast value. The values can range from −100 to +100.

Drag the contrast slider in Figure 6.47 right to approximately +75. Observe the increase in detail in the image. The additional contrast has increased the separation between the highlights (light areas) and shadows (dark areas) of the image. This works very well on this image, because there are very few midtones to worry about.

Let's experiment a little. Click on the Cancel button in the dialog box. Choose the Image > Adjust > Levels command. Observe the histogram in the Levels dialog box; the image has nothing but midtones – no highlights or shadows. Drag the sliders in to the first level of pixels on both ends of the input as shown in Figure 6.48. If you look closely, this is just about the same result that you received with the brightness/contrast adjustments.

Click OK in the dialog box to apply the adjustments.

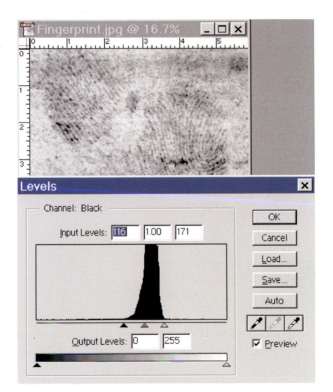

Figure 6.48

SHARPENING THE IMAGE

Now it's time to discuss sharpening the image. The image has been sized, cropped if necessary, and enhanced. All digital images lose a little sharpness in the electronic process. Scanned images usually require more sharpening than images from a digital camera, but running a test on your images is the best way to determine how much sharpening an image requires.

Unsharp masking, or USM, is a traditional film compositing technique used to sharpen edges in an image. The unsharp mask filter corrects blurring introduced during photographing, scanning, resampling, or printing. It is useful for images intended both for print and online.

Choose the Filter > Sharpen > Unsharp Mask command. The Unsharp Mask dialog box will open. Numbers represent unsharp mask amount, radius, and threshold (Figure 6.49). The unsharp mask filter locates pixels that differ from surrounding pixels by the threshold you specify and increases the pixels' contrast by the amount you specify. In addition, you specify the radius of the region to which each pixel is compared.

The effects of the unsharp mask filter are far more pronounced on-screen than in high-resolution output. If your final destination is print, experiment to determine what dialog box settings work best for your image.

Figure 6.49

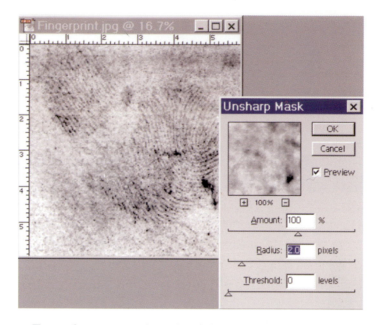

To set the amount, drag the slider or enter a value to determine how much to increase the contrast of pixels. For high-resolution printed images, an amount between 100% and 200% is recommended. Set this image to 100%.

To set the radius, drag the slider or enter a value to determine the number of pixels surrounding the edge pixels that affects the sharpening. For high-resolution images, a radius between 1 and 2 is recommended. A lower value sharpens only the edge pixels, whereas a higher value sharpens a wider band of pixels. This effect is much less noticeable in print than on-screen, because a 2-pixel radius represents a smaller area in a high-resolution printed image. Set this image to a radius of 1.

To set the threshold, drag the slider or enter a value to determine how different the sharpened pixels must be from the surrounding area before they are considered edge pixels and sharpened by the filter. To avoid introducing noise (in images with flesh tones, for example), experiment with threshold values between 2 and 20. The default threshold value (0) sharpens all pixels in the image. When sharpening fingerprints set the threshold to 0 so all pixels in the image will be sharpened. The default works quite well on all images.

If applying the unsharp mask filter makes already bright colors appear overly saturated, convert the image to Lab mode and apply the filter to the L channel only. This technique sharpens the image without affecting any of the color components.

Click OK to apply the unsharp mask.

Choose the File > Save As command. Keep the same file name and save it in Your Folder.

You have made all the basic adjustments to an image, size, crop, enhance, and sharpen. You are ready to print your images. Chances are that these are all the adjustments that will be necessary on the majority of the images you encounter.

Close the image.

CURVES ADJUSTMENT

The most sophisticated tonal adjustment tool in Photoshop is the Curves adjustments. Like Levels, Curves lets you adjust the tonal range of an image. However, instead of making the adjustments using just three variables (highlights, shadows, and midtones), you can adjust any point along the 0–255 scale while keeping up to 15 other values constant. Curves lets you make precise adjustments to one area of the tonal range while controlling the effect on the others. For RGB images, Curves displays brightness values from 0 to 255, with shadows (0) on the l to 255 scale.

Choose the File > Open > Fingerprint.jpg from the CD.

Choose the Image > Adjust > Curves command. The Curves dialog box will open (Figure 6.50).

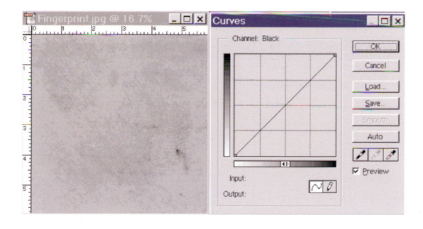

Figure 6.50

First you must sample the image and click the part of the curve you want to adjust. A set of direction arrows marks the pixel's position on the graph, and the input and output values appear at the bottom of the dialog box.

Move onto the image with the eyedropper and locate the darkest part of the image. Sample it by clicking the eyedropper and observing the input value in the dialog box. You want to use a black that still has detail, in other words less that 100%. There is a 95% black in this image. To set the 95% anchor on the curve, move up the straight line until 95% appears in the input box as shown in Figure 6.51. Click the mouse button to set the anchor point.

Figure 6.51

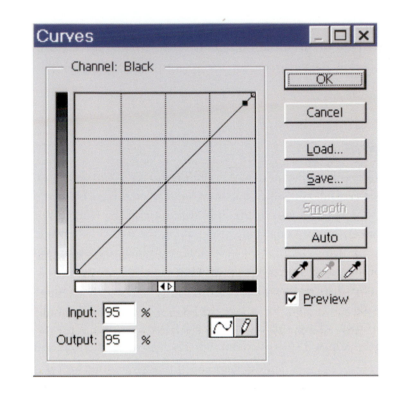

Figure 6.52

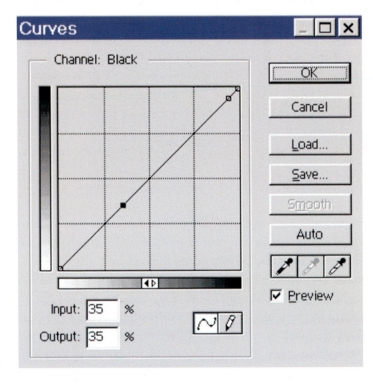

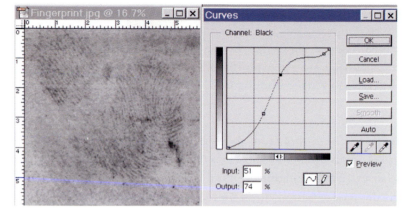

Figure 6.53

Now you must locate the white point in the image. Again we need a white area with detail; never choose 0%. This will not be a problem in this image, because there is no true white. About as close as you will get is 30–35% as shown in Figure 6.52. Click on the straight line to set the anchor point. There should be two anchor points now, one for the highlight (light areas) and one for the shadows (dark areas).

Place the mouse approximately halfway between the anchor points on the straight line. Drag the curve up until the image looks as you want it (Figure 6.53). You should have about the same results as with the Levels and Contrast/Brightness adjustments.

The gamma of any curve is its slope, expressed as the ratio of the logs of the output to input values. For example, a gamma value of 1.0 equals an output-to-input ratio of 1:1. Moving the midpoint of the curve up (in an RGB readout) lowers the gamma value (that is, to less than 1.0); moving the midpoint down raises the gamma value. (This is the same as moving the gamma levels slider to the right to lower the gamma level, left to raise it.) The Auto levels button and the eyedroppers work the same as they did in the Levels dialog box and make the same adjustments.

Apply the unsharp mask filter as you did earlier. Choose File > Save As; rename the image Fingerprint_C, and save the image to Your Folder.

Close the image.

SELECTION TOOLS

Photoshop has various selection tools available. There are a number of reasons you would use a selection tool, but the main reason is that you want to work on the area within the selection.

There are basically two types of adjustments you can make in Photoshop: global adjustments, which means you are adjusting the entire image, and selected adjustments, which means you are only applying adjustments to the area within a selection marquee.

Any shape or size can be selected within an image if you use the correct tool and have a little patience. We will explore each tool and see how it functions.

Choose File > Open > Fingerprint.jpg from the CD.

Lasso tool

 Select the lasso tool from the toolbox. The lasso selection tool is used to make freehand selections. We will select only the fingerprint so we can adjust the tones in it only. Place the mouse outside the fingerprint, hold the mouse button down and drag a selection marquee around the fingerprint portion of the image (Figure 6.54). Ensure that you complete the selection before you release the mouse button or it will put a straight line back to where you began the selection. Release the mouse button. There should be a marquee around the entire fingerprint. You cannot adjust the size of the marquee, but you can move inside the marquee and the cursor will become an arrow. Hold down the mouse button to move the marquee. If you have to start over, choose the Select > Deselect command and create the marquee again. To observe how selected adjustments work, choose the Image > Adjust > Brightness/Contrast command as in Figure 6.55. Apply a +75 to the contrast and observe the image.

Figure 6.54

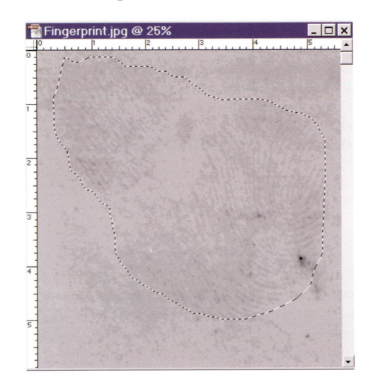

Figure 6.55

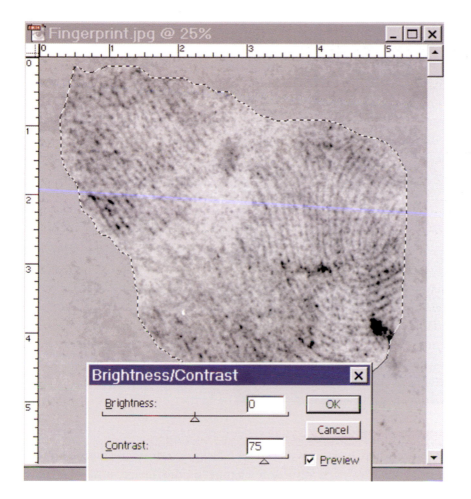

Click Cancel in the dialog box. Close the image without saving it.

Note: You can change the lasso cursor to a crosshair cursor by pressing the Caps Lock key. The crosshair is more accurate when dragging the selection marquee.

Choose Image > Open > Image_1.jpg from the CD.

Polygon lasso tool

 Hold the mouse button down and drag to select the polygon lasso tool from the toolbox.

The main difference between the polygon lasso and the lasso tool is that it selects straight lines. Let's select the dark wooden portion of the building as shown in Figure 6.56. Position the cursor at the beginning of a straight-line portion in the image.

Click and release the mouse button. Move the cursor to the end of the

Figure 6.56

straight line or where it changes directions. Click and release the mouse button. Continue to click at each direction change in the image until you have completed the selection. When the cursor is positioned back at the first point you created, a circle appears, indicating that you can now close the polygon. Click the mouse button on the beginning point to close the polygon and a selection marquee will appear.

Use the Brightness/Contrast tool to lighten the selected area.

Close the image without saving it.

Choose File > Open > Fingerprint_2.jpg from the CD.

Magnetic lasso tool

 Select the magnetic lasso tool from the toolbox. The magnetic lasso tool works by setting first a fastening point in the image. Fastening points anchor the selection border in place.

You will select just the finger in Figure 6.57. To draw a freehand segment, move the pointer along the edge you want to trace. (You can also drag with the mouse button depressed.) The most recent segment of the selection border remains active. As you move the pointer, the active segment snaps to the

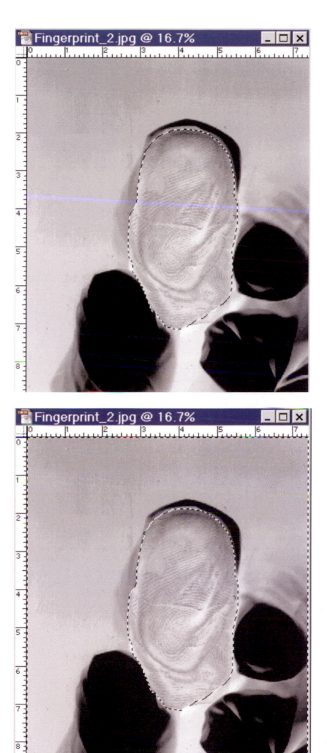

Figure 6.57

Figure 6.58

strongest edge in the image. Periodically, the magnetic lasso tool adds fastening points to the selection border to anchor previous segments.

If the border doesn't snap to the desired edge, click once to add a fastening point manually. Continue to trace the edge and add fastening points as needed. Complete the selection so you have a marquee around the finger, because we are going to remove the background from the image.

First ensure that the default colors are set in the toolbox (Figure 6.58). The foreground must be black and the background white. Click on the default color icon if necessary.

Choose the Select > Inverse command. This will inverse the selection and everything will be selected in the image but the finger.

Choose the Edit > Clear command. The background is cleared and becomes white (Figure 6.59).

Choose the File > Save As command. Name the image Fingerprint_2B and save it in Your Folder.

Close the image.

Figure 6.59

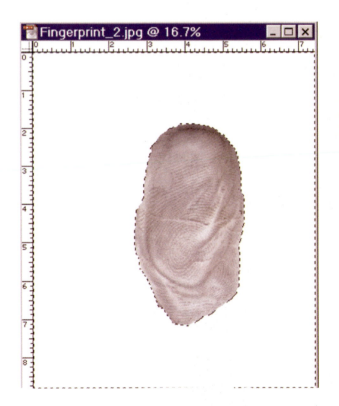

Rectangular marquee

 The rectangular marquee is used to select any rectangular area in an image. Choose File > Open > Image_2.jpg from the CD.

Hold down the mouse button and drag to select the rectangular marquee tool in the toolbox.

Select the doorjamb on the left side of the image so you can enhance it as shown in Figure 6.60. Position the cursor at one corner of the rectangular area you want to select. Drag diagonally to create a rectangular marquee around the area you want to select. Release the mouse button.

Use the levels tool and the unsharp mask to increase the detail in the image as shown in Figure 6.61.

Figure 6.60

Figure 6.61

To constrain the selection and make a square, hold the Shift key while dragging the marquee – ensure that you release the mouse button before the Shift key. To select from the center of the area hold the Alt key and drag to draw a rectangle from the center out – ensure that you release the mouse button before the Alt key.

Holding the Alt and Shift keys will maintain a rectangle from the center.

Note: All marquees may be moved by placing the cursor inside the selected area and dragging. No adjustments to the size of the marquee can be made after it appears, you must remove it and start over.

Close the image without saving it.

Choose File > Open > Image_3.jpg from the CD.

Elliptical marquee tool

 Select the elliptical marquee tool from the toolbox. Position the cursor outside the object you want to select as shown in Figure 6.62. Drag diagonally to create an elliptical marquee around the area you want to select. Release the mouse button.

To constrain to a circle hold the Shift key while dragging the marquee to make a perfect circle – ensure that you release the mouse button before the Shift key. To select an area from the center hold the Alt key down while dragging. Holding the Alt and Shift keys will maintain a circle from the center.

Close the image without saving it.

Figure 6.62

Magic wand tool

The magic wand is probably the most powerful selection tool in Photoshop, if you understand how to use it correctly. The magic wand tool lets you select a consistently colored area (for example, a red shirt) without having to trace its outline. You specify the color range, or tolerance, for the magic wand tool's selection.

Note: You cannot use the magic wand tool on an image in Bitmap mode.

For tolerance, enter a value in pixels, ranging from 0 to 255. Enter a low value to select colors very similar to the pixel you click or a higher value to select a broader range of colors.

To define a smooth edge, select Anti-aliased. Choose File > Open > Fingerprint_2B from Your Folder.

Double-click the magic wand tool in the toolbox. The magic wand options appear in the options palette. Enter a value of 1 in the Tolerance field. Check Anti-Aliased and Contiguous as shown in Figure 6.63.

Move the cursor onto the white area of the image and click. A selection marquee will appear, selecting the entire background. The tolerance was set so low that all the magic wand sampled was the white area.

Choose the Select > Inverse command. Now the fingerprint is selected so you can make adjustments.

Note: The magic wand tool works by reading the brightness of the pixel you click on, and comparing the brightness values of surrounding pixels. If an adjacent pixel is within the tolerance you have chosen (lighter or darker) it will be selected.

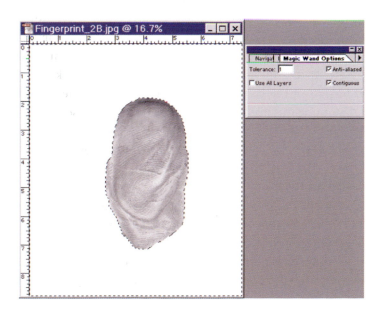

Figure 6.63

Figure 6.64

This was a simple selection; now let's try a more difficult selection.

Close the image without saving it.

Choose File > Open > Image_1 from Your Folder.

Zoom in on the graffiti as shown in Figure 6.64. Place the options palette on the image. Set the tolerance to 30. Click in a red portion of the playing cards as shown. This will not select all the red; you must use a technique of multiple selection. After you have made the original selection, hold the Shift key and click in another red section of the cards. This technique will add the second selection to the first. Continue this process until you have all the red selected as shown in the illustration.

Now you can use the levels or contrast/brightness tools to enhance the selected area.

Close the image without saving it.

It would be a perfect world if every selection we made were correct, but most of the time we need to make adjustments or add/subtract from the selection.

Choose File > Open > Image_4 from the CD.

Double-click on the magic wand tool, set the tolerance to 25. We're going to select the white tile and cloth by the door with the bloodstains so we can enhance the detail.

Figure 6.65

Click the mouse button on the white cloth as indicated in Figure 6.65. It will select part of the white cloth and probably some of the doorjamb. Don't worry about it, we'll remove it later. Now you will multiple select with the magic wand as you did before with the graffiti. Hold the Shift key and continue sampling additional areas that you want to select. Continue sampling, holding the Shift key, and select as much of the area as possible. Don't worry about getting too much or too little of the area. Now you will remove the doorjamb or any other area you don't want in your selection. Select the lasso tool outside the selection approximately as shown in Figure 6.66. To subtract from a selection, the cursor must be outside the selection area.

Hold down the Alt key and drag a marquee around the selected doorjamb you want to remove. Release the mouse button only when the cursor is outside the original selection area where you began. You must be accurate, but this should remove the doorjamb from the selection as in Figure 6.67. Continue this process if there are any other areas to remove.

Now let's add the areas inside the selection. The cursor must be inside the selected area.

Hold down the Shift key. Hold down the mouse button and drag a marquee around the areas you want to add to the selection – ensure that you come completely back inside the selection to where you began before releasing the mouse

Figure 6.66

Figure 6.67

Figure 6.68

button. The area that you encircled will be added to the original selection. Continue this until all the areas are added as shown in Figure 6.68.

Close the image, do not save.

TRANSFORMING SELECTIONS

All selections can also be transformed in Photoshop. The Free Transform command lets you use the Scale, Rotate, Skew, Distort, and Perspective commands without having to select them from the menu. Understand that only Scale and Rotate would normally be used. You're not normally going to skew, distort, or change the perspective of any evidentiary photographs.

To access the various transformation modes, you use different shortcut keys as you drag the handles of the transform-bounding box.

Choose File > Open > Fingerprint_1 from the CD. Select the lasso tool and tightly place a marquee around the fingerprint as shown in Figure 6.69.

To freely transform an area of the image, it must be selected prior to using the Transform command.

Transform as follows: choose the Edit > Free Transform command. A bounding border appears around the selection with handles. Experiment using

Figure 6.69

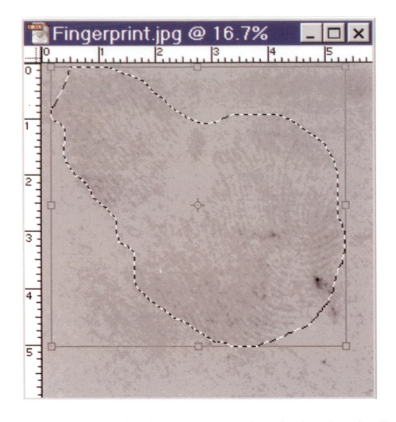

the controls listed below to transform the selection. Pressing Enter applies the transformation. To cancel the transformation, press Esc.

To move, position the pointer inside the bounding border (it turns into a black arrowhead), and drag.

To scale, drag a handle. Press Shift as you drag a corner handle to scale proportionately. When positioned over a handle, the pointer turns into a double arrow.

To rotate, move the pointer outside of the bounding border (it turns into a curved, two-sided arrow), and then drag. Press Shift to constrain the rotation to 15° increments.

To distort freely, press Ctrl and drag a handle. To distort relative to the center point of the bounding border, press Alt and drag a handle.

To skew, press Ctrl+Shift and drag a side handle. When positioned over a side handle, the pointer turns into a white arrowhead with a small double arrow.

To apply perspective, press Ctrl+Alt+Shift and drag a corner handle. When positioned over a corner handle, the pointer turns into a gray arrowhead.

Selections can also be transformed numerically. The Numeric command lets you scale, rotate, skew, or move part of the image, a layer, path, or selection border precisely, by entering specific numeric values for the transformation.

If after experimenting with the Free Transform command, the image is unusable, revert so you have a clean image. Select the fingerprint with the lasso tool.

Choose Edit > Transform > Numeric.

Experiment with the numeric controls (Figure 6.70) as described below.

To move, select Position. Choose the unit of measurement from the menus; then enter a horizontal distance in the X text box and a vertical distance in the Y text box. To move relative to the existing image, layer, or path, select Relative. Otherwise, the movement is relative to the top left corner of the image.

To resize, select Scale. Choose the unit of measurement from the menus; then enter values for the Width and Height. Select Constrain Proportions to scale proportionally.

To skew, select Skew. Then enter values for the Horizontal and Vertical angles of slant.

To rotate, select Rotate. Then enter a value for Angle or drag the radius inside the circle to the desired angle of rotation.

Click OK to complete any transformations. Close the image, do not save.

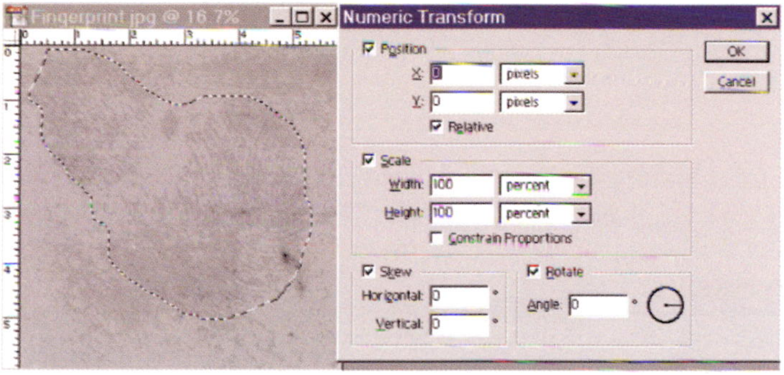

Figure 6.70

COLOR IN PHOTOSHOP

Before we discuss the colors in-depth we need to understand what color gamut is and what it means.

COLOR GAMUT

The gamut is the range of colors that can be displayed or printed in a color system. A color that can be displayed in RGB or HSB models may be out-of-gamut, and therefore unprintable, for your CMYK setting. Photoshop automatically brings all colors into gamut when you convert an image to CMYK. But you might want to identify the out-of-gamut colors in an image or correct them manually before converting to CMYK.

In RGB mode, you can identify out-of-gamut colors in the following ways.

In the Info palette, an exclamation point appears next to the CMYK values whenever you move the pointer over an out-of-gamut color. In both the color picker and the color palette, an alert triangle appears and the closest CMYK equivalent is displayed whenever you select an out-of-gamut color. To select the CMYK equivalent, click the triangle or the color patch.

OUT-OF-GAMUT COLOR INDICATOR

You can also quickly identify all out-of-gamut colors in an RGB image by using the Gamut Warning command. This command highlights all pixels that are out-of-gamut in the image.

To turn on and off the highlighting of out-of-gamut colors: choose View > Gamut Warning. To change the gamut warning color: Choose File > Preferences > Transparency & Gamut. Under Gamut Warning, click the color box to display the color picker; then choose a new warning color. Enter a value in the Opacity text box. Values can range from 0 to 100%. Use this setting to reveal more or less of the underlying image through the warning color. Then click OK. Choose File > Open > Image_1.jpg from Your Folder.

COLOR PALETTE

Drag the color palette next to your image as shown in Figure 6.71. Notice that the foreground and background colors displayed in the toolbox also are shown

Figure 6.71

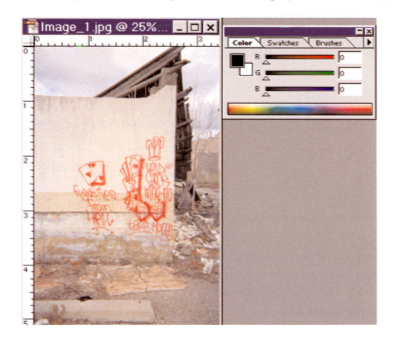

in the color palette. Using the sliders in the color palette, you can edit the foreground and background colors according to several different color models. You can also choose a background or foreground color from the spectrum of colors displayed in the color bar. Click in the center of the color bar. The foreground color changed to blue. Observe the out-of-gamut indicator that appears in the color palette in Figure 6.72. Click on the box next to the indicator to select the closest color that will print.

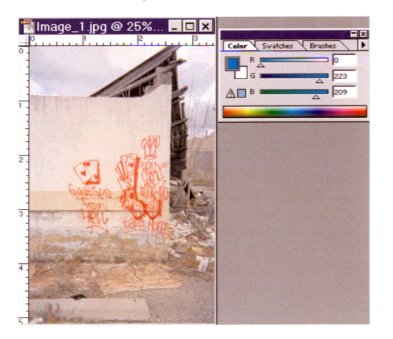

Figure 6.72

INFO PALETTE

Drag the info palette over next to the image as shown in Figure 6.73. Move the cursor over the image and watch the CMYK values. If you move into the red graffiti the exclamation point will appear, indicating it is out-of-gamut. All colors in an image can be sampled and checked to see if they are out-of-gamut. Clicking in the color bar set the foreground color. Hold down the Alt key to set the background color.

While you are using the info palette, observe that it also displays other information. When using the marquee tool, it displays the *x* and *y* coordinates of your starting position and the width (W) and height (H) of the marquee as you drag. When you use the crop tool or zoom tool, it displays the width (W) and height (H) of the marquee as you drag. The palette also shows the angle of rotation of the crop marquee. When you use the line tool, pen tool, or gradient tool or when you move a selection, the it displays the *X* and *Y* coordinates of your

Figure 6.73

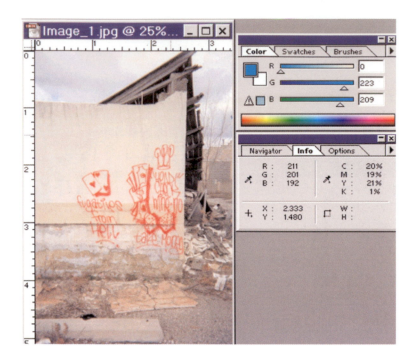

starting position, the change in *X* (DX), the change in *Y* (DY), the angle (A), and the distance (D) as you drag.

When you use a two-dimensional transformation command, it displays the percentage change in width (W) and height (H), the angle of rotation (A), and the angle of horizontal skew (H) or vertical skew (V).

When you use any color adjustment dialog box (for example, Curves), it displays the before and after color values of the pixels beneath the pointer and beneath color samplers. Experiment with the various tools and observe the info palette. It is recommended that the info palette be open any time you are working on an image.

The color picker can also be used to edit the foreground or background.

COLOR PICKER

Click on the foreground color indicator in the toolbox. The foreground color picker opens. The dialog box shows you the present foreground color selected and has a color field for selecting a new color by clicking in it. There is also a color bar that you can select from any color range.

Click in the top right corner of the color field in Figure 6.74. A pyramid appears to the right indicating that this color is out-of-gamut. Notice that the color mixtures are also indicated in RGB, CMYK, HSB, and Lab.

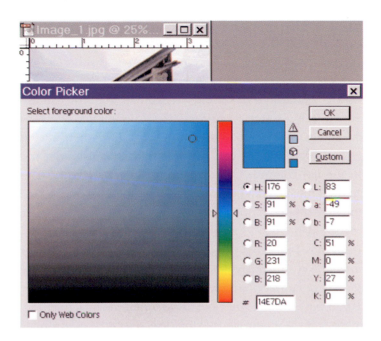

Figure 6.74

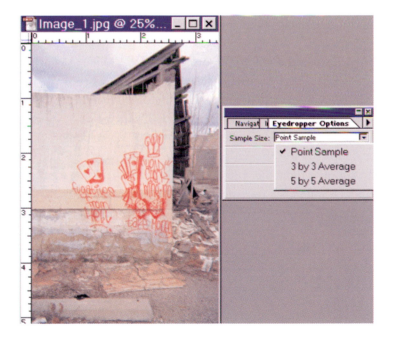

Figure 6.75

Experiment by clicking at different places in the color field, sliding the slider up and down, and clicking the radio buttons next to the color controls.

Note: The background color is selected in the same manner.

Click OK when you have selected a foreground color.

While we are discussing colors, let's look at the eyedropper tool. You can use the color palette and the info palette to preview the color values of pixels affected by color adjustments. When you work with a color adjustment dialog box, the info palette displays two sets of color values for the pixels under the pointer. The value in the left column is the original pixel's color value. The value in the right column is the color value after the adjustment is made.

Double-click on the eyedropper tool to select it and open the eyedropper options palette. Move the palette next to the image as shown in Figure 6.75.

You can view the color of a single area using the eyedropper tool, or you can use up to four color samplers to display color information for one or more locations in the image. These samplers are saved in the image, so you can refer to them repeatedly as you work, even if you close and reopen the image.

To view color samplers and color sampler information, select the eyedropper or color sampler tool, or open a color adjustment dialog box.

To display or hide color sampler information in the info palette, choose Show Color Samplers or Hide Color Samplers from the info palette menu.

To place a color sampler, select the color sampler tool and click where you want to place it.

To set the area measured by the color sampler tool, select the color sampler tool.

Choose what to sample (Figure 6.75): select Point Sample to read the value of a single pixel. Select 3 by 3 Average to read the average value of a 3×3 pixel area. Select 5 by 5 Average to read the average value of a 5×5 pixel area.

To delete a color sampler, select the color sampler tool.

Delete the color sampler, drag the sampler out of the document window. Hold down Alt (Windows) and click the sampler.

To move a color sampler, select the color sampler tool and drag it to the new location.

(Note: The sampler does not move to the new location until you release the mouse button.)

To change the color space in which a color sampler displays values, move the pointer onto the color sampler icon in the info palette. Hold down the mouse button and choose a color space from the menu.

To use the info palette to preview color changes, choose Window > Show Info. Place one or more color samplers if desired. As with the eyedropper tool, the color sampler reads a 1×1, 3×3, or 5×5 pixel area.

Open a color adjustment dialog box. Move the pointer over an area of the image you want to examine. The eyedropper reads the value of a single screen pixel, a 3×3 screen pixel area, or a 5×5 screen pixel area, depending on the sample size option chosen in the eyedropper options palette. (*Note:* Opening a color adjustment dialog box activates the eyedropper tool outside the dialog box. However, you still have access to the scroll controls and to the hand and zoom tools when using keyboard modifier keys.)

Make any color adjustments, and before applying them, view the before and after color values in the Info palette.

To use the color palette to preview color values, choose Window > Show Color. Select the eyedropper tool. Open a color adjustment dialog box. Click the pixel you want to preview in the image.

Close the image, do not save it.

NEW CANVAS

Before beginning the next session on drawing tools, let's see how to build a new canvas. The canvas is just like painting a picture; this is what you begin with and what you will build the image on. The canvas also lets you add or remove workspace around an existing image. You can crop an image by decreasing the canvas area. Added canvas appears in the same color or transparency as the background.

To build a new canvas, select the File > New command (Ctrl + N). The New dialog box will open (Figure 6.76). Name the new canvas Sample and enter the

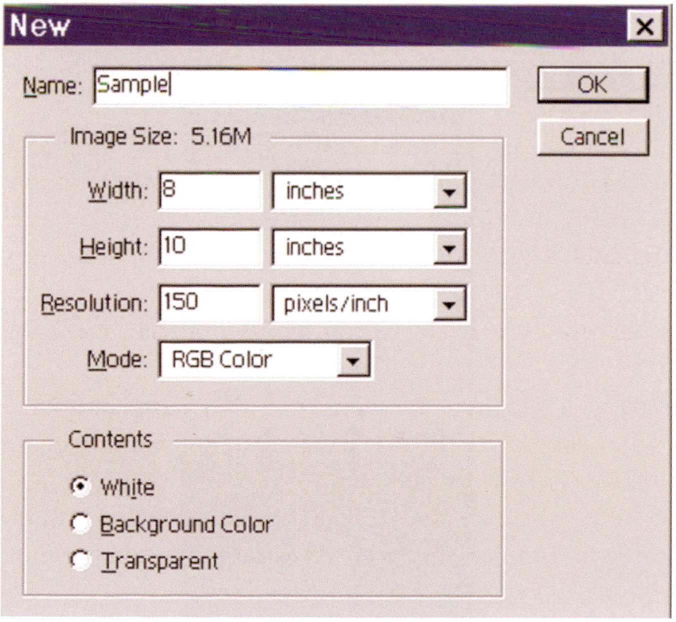

Figure 6.76

values shown. Click OK. The new canvas is very handy for building composites to show in court. We will address that subject later. Right now we will use the canvas to demonstrate the drawing tools.

DRAWING TOOLS

Photoshop has a number of drawing tools; some are used for forensics and some you will never use in this field. The main tools include the paintbrush, pencil, line, paint bucket, and gradient tool. We will discuss them briefly and show how they are used.

PAINTBRUSH

 Choose the paintbrush tool from the toolbox. Double-click on it to open the paintbrush options palette. Open the brushes palette. Drag both of them next to the image as shown in Figure 6.77.

The options and the brushes palette will be used to set up the paintbrush. Basically the options palette controls the opacity and the brushes palette controls the size and whether the brush is hard-edged or feathered.

Click on a hard-edged brush in the center of the top row. Drag the paintbrush across the top of the canvas. Click on a soft-edged brush in the center of the second row. Drag the paintbrush across under the last brush stroke on the canvas.

The centers of the hard and soft-edged strokes are similar, but the edges of the soft-edged stroke fades to the color of the pixels you painted over (Figure 6.78).

Figure 6.77

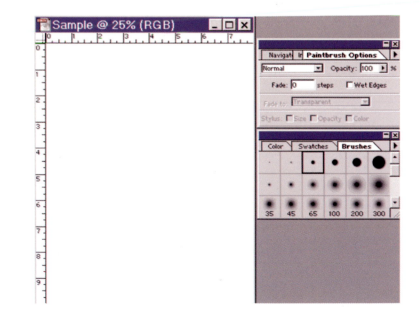

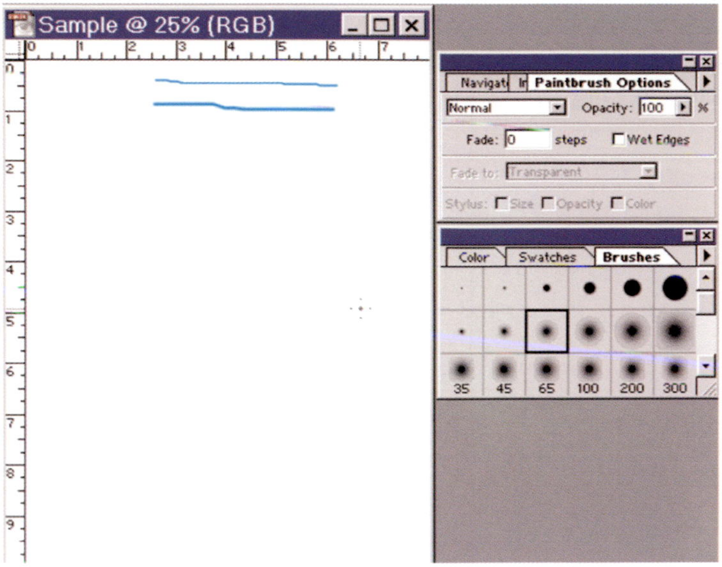

Figure 6.78

The paint opacity can be adjusted prior to painting. Drag the opacity slider in the paintbrush options palette to 40%. Using the paintbrush tool, paint another line under the existing ones. Changing opacity allows you to see the existing colors of the image or canvas through the paintbrush.

The paintbrush is more of an artistic tool, but has some applications using the masking technique in Photoshop.

PENCIL

 Choose the pencil tool from the toolbox. Notice that the brushes palette changes; all the brushes become hard-edged (Figure 6.79).

Choose a center brush in the second row. Write the word "Test" in the middle of the image. This is a very simple freehand tool, but will rarely be used in forensics.

LINE TOOL

Choose the line tool from the toolbox. The line tool has many uses when building composites for the courtroom. Observe the options palette. The width of the line is set with the weight value, not the brushes palette. Enter a value of 12 and drag a line under the word test. Holding the Shift key will maintain a straight line or 45° increments.

Check Start agaianst Arrowheads and drag another straight line as shown in Figure 6.80.

The paint bucket and gradient tools are both used for filling selected areas of

Figure 6.79

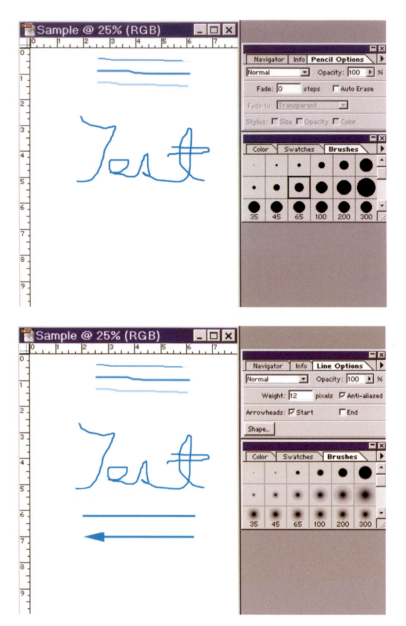

Figure 6.80

the image, but they can also be used to create masks in Photoshop, as we will discuss later. Select the Rectangle tool and make a selection as shown in Figure 6.81. Choose the Edit > Fill command.

PAINT BUCKET

 Options for filling the selected area vary, but for the masking technique, use the foreground color. Leave blending at normal, blending modes will be discussed in detail later in the chapter. Leave

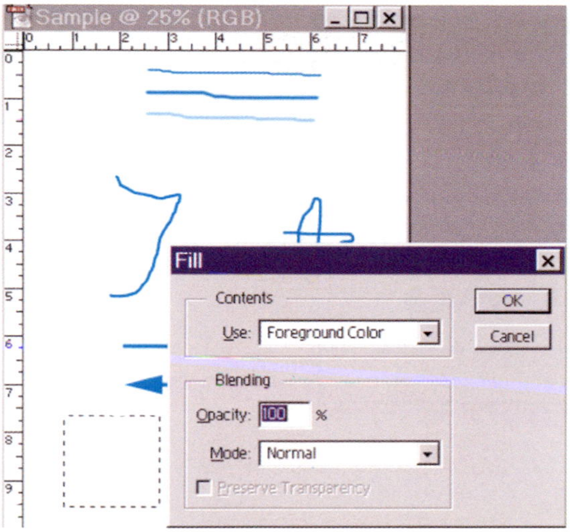

Figure 6.81

opacity at 100%, click OK and the selected area will be filled with the foreground color. Any selected area, including the complete image, can be filled with the paint bucket.

GRADIENT TOOL

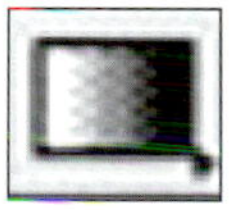

The gradient tool works much the same way as the paint bucket, except that you get to set the beginning and ending color. Select the gradient tool from the toolbox. You've probably already noticed that there are six of them. The only difference is the way they fill an area. Remove any selections from the image. Select the linear gradient tool. Double-click on it to open the options palette. The options palette has many of the same controls as the paint bucket. Leave it at normal with opacity of 100%. Click the drop-down arrow for the gradient setting and observe the many options. Select Foreground to Background as shown in Figure 6.82. Move the mouse onto the image and begin at the bottom center of the image as shown by the arrow in Figure 6.83.

Hold down the mouse button and drag to the top of the image. Release the mouse button. The image will be filled with a gradient, starting with the foreground color at the bottom and ending with the background color at the top of the image as shown in Figure 6.84.

This same technique will be used later in an exercise to demonstrate using a mask with a blending mode.

Choose the File > Save As command and place this image in Your Folder. We will use it later for another exercise.

Close the Image.

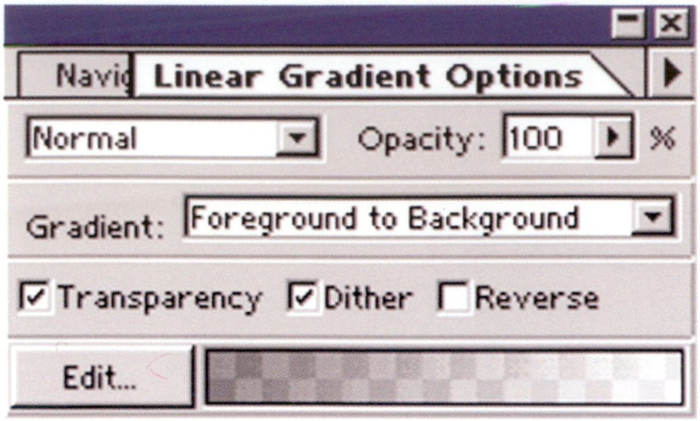

Figure 6.82

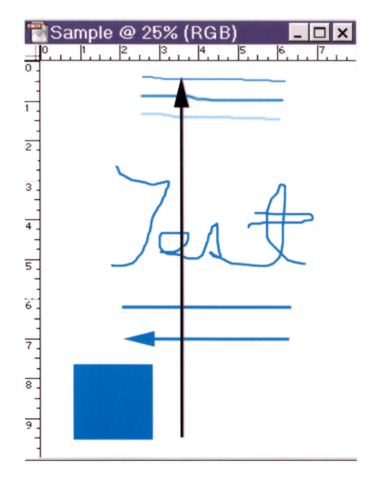

Figure 6.83

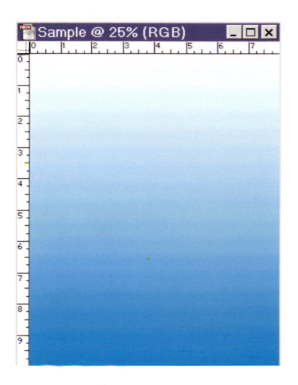

Figure 6.84

UNDO COMMAND

Most people using a program like Photoshop have a favorite tool and many times this is the Undo command. It sure can get you out of a lot of trouble if you have made a mistake. In the past, many programs only let you undo the last action applied to the image. Photoshop, in addition to the Undo command in the Edit menu, maintains a history of the actions in a history palette.

HISTORY PALETTE

The history palette lets you jump to any recent state of the image created during the current working session. Each time you apply a change to an image, the new state of that image is added to the palette. For example, if you select paint and rotate part of an image, each of those states is listed separately in the palette. You can then select any of these states, and the image will revert to how it looked when that change was first applied. You can then work from that state.

The following guide can help you with the history palette:

■ Program-wide changes, such as changes to palettes, color settings, actions, and preferences, are not changes to a particular image and so are not added to the history palette.

- By default, the history palette lists the previous 20 states. Older states are automatically deleted to free more memory for Photoshop.
- To keep a particular state throughout your work session, make a snapshot of the state.
- Once you close and reopen the document, all states and snapshots from the last working session are cleared from the palette.
- By default, a snapshot of the initial state of the document is displayed at the top of the palette.
- States are added from the top down. That is, the oldest state is at the top of the list, the most recent one at the bottom.
- By default, deleting a state deletes that state and those that came after it. If you choose the Allow Non-Linear History option, deleting a state deletes just that state.

Ensure that the history palette is showing, if not, choose the Window > Show History command.

Choose File > Open > Image_1.jpg from the CD.

Apply the basic enhancements on your image as you did earlier in the chapter:

- Size the image
- Enhance the image with the levels tool
- Apply an unsharp mask

Figure 6.85

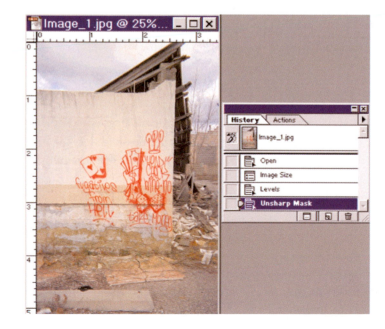

Notice each state of the image as it is recorded in the History palette (Figure 6.85).

Click on the Open state in the history palette. All other states are dimmed and the image appears as it was when it was opened, as shown in Figure 6.86.

Click each individual state as you move down the list – the history palette shows the image in each individual state. Click on the Levels state. Click on the drop-down arrow in the upper-right corner to open the menu.

Choose the delete command from the menu (Figure 6.87). All states are deleted from that point on in the process. The history brush tool works in conjunction with the history palette and lets you paint a copy of one state or snapshot of an image into the current image window. This tool makes a copy, or sample, of the image and then paints with it.

Figure 6.86

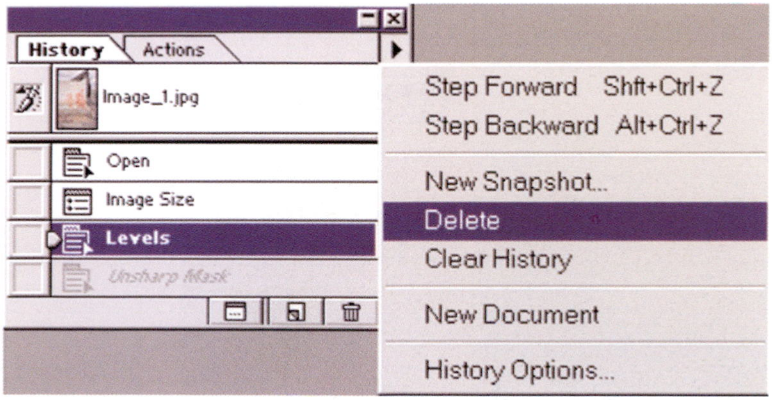

Figure 6.87

For example, you might make a snapshot of a change you made with a painting tool or filter. After undoing the change to the image, you could use the history brush tool to apply the change selectively to areas of the image. Unless you select a merged snapshot, the history brush tool paints from a layer in the selected state to the same layer in another state.

Experiment with the history brush tool.

Close the image, do not save it. Remember that the history is eliminated when an image is closed.

Choose File > Open > Sample.jpg from Your Folder. Let's see how the eraser tool works.

ERASER TOOL

 The eraser tool changes pixels in the image as you drag through them. If you're working in the background, or in any other layer with Preserve Transparency on, the pixels change to the background color. Otherwise, transparency replaces the color. You can also use the eraser to return the affected area to a state selected in the history palette.

Erasing pixels from a layer exposes the underlying layer. Set the default colors in the toolbox by clicking on the default color icon. Double-click the eraser tool to display its options palette. Choose the tool type you want to use as an eraser – paintbrush, airbrush, pencil, or block – set it to paintbrush. To cycle through the eraser tool types, hold down Alt and click the eraser tool, or hold down Shift and press E.

Figure 6.88

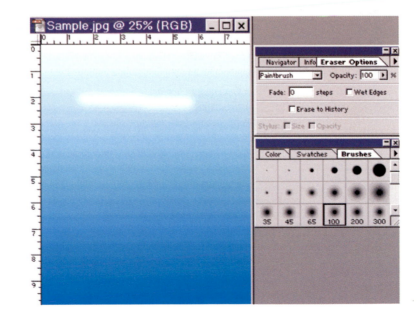

Set the opacity to 100%, fade-out rate to 0, wet edges and erase to history off. Select a brush size from the brushes palette.

Drag the mouse across the image. The image is erased to the background color as shown in Figure 6.88.

To erase to a saved state or snapshot of the image, select Erase to History. To use the eraser tool in Erase to History mode, hold down Alt as you drag in the image.

Close the image, do not save.

RUBBER STAMP TOOL

The rubber stamp tool is commonly referred to as the "clone" tool and is probably the biggest manipulation tool in Adobe Photoshop. It is not recommended for use on any evidentiary photographs, but you should understand how to use it and what effect it has on the image.

Choose File > Open > Image_4.jpg from the CD. Zoom in on the floor area that contains the ashtray as shown in Figure 6.89.

Choose the rubber stamp tool from the toolbox and open the brush palette. Choose the fifth brush in the first row. This determines how large the area will be that is sampled and cloned. First you must sample the portion of the image you want to clone from. Hold down the Alt key and click the mouse button on the carpet area to the right of the ashtray. Release the Alt key and the mouse button.

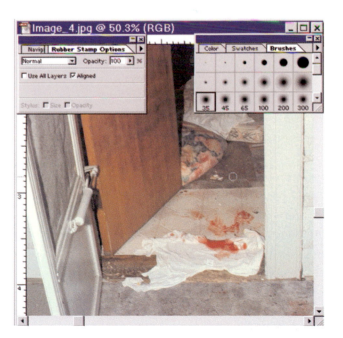

Figure 6.89

Move the rubber stamp to the left onto the edge of the ashtray. Hold down the mouse button – notice a crosshair appears in the area of the image where you sampled. This is the portion of the image that will be reproduced in the area you clone to. Hold down the mouse button and drag left and right and up and down, duplicating the area that you sampled. This will replace the ashtray pixels with pixels from the carpet as shown in Figure 6.90.

Note: Watch the crosshair closely while you are cloning.

Release the mouse button.

This tool is dangerous; it replaces pixels in the image and as you can see completely alters the image and it cannot be detected. The ashtray that was removed could just as easily have been a weapon.

Close the image, do not save.

Figure 6.90

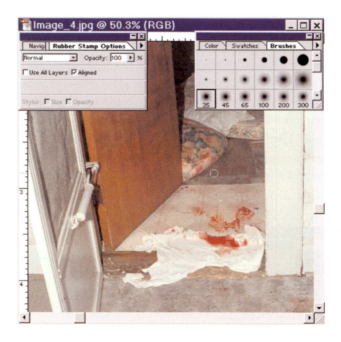

ACTIONS PALETTE

The actions palette in Photoshop can make your job easier and be a great time saver. It lets you automate tasks by grouping a series of commands into a single action. For example, you can create an action that applies a series of filters to reproduce a given effect on an image. Actions can be grouped into sets to help you better organize your actions. You can use an action on a single file or on a batch (multiple files in the same folder).

It works just like a tape recorder. You use the actions palette to record, play, edit, and delete actions. It also lets you save, load, replace and print actions.

Ensure that the actions palette is showing; if not, choose Window > Show

Actions. You can display actions in the actions palette in either list view or button view. In list view, sets can be expanded to display actions, which in turn can be expanded to display individual commands. Commands can then be expanded to display recorded values.

To display actions as buttons, choose Button Mode from the actions palette menu. Choose Button Mode again to return to the list view.

To expand and collapse sets and commands, click the triangle to the left of the set or command in the actions palette.

Recording actions

The actions palette will always open at default with the last set of actions that were recorded.

The first action we will record is a sizing action. It is very useful when importing digital images from many of the point-and-shoot cameras. Most of those images come in very large sizes and low resolution. Each must be resized for printing. Before beginning, build a new folder in Your Folder. Name it Sizing.

Choose File > Sizing Folder > Size_1.jpg from the CD. Drag the action palette out next to the image. The first step is to clear the actions in the palette. Click on the arrow in the upper right corner of the action palette and a drop-down menu will appear as shown in Figure 6.91. Select Clear Actions and the information in the action palette will be deleted. If these actions were saved as a file, it does not delete the file.

Using the same drop-down menu, click on the New Action command at the top of the menu. The New Action dialog box will open (Figure 6.92). Enter a name for the new action.

Figure 6.91

Figure 6.92

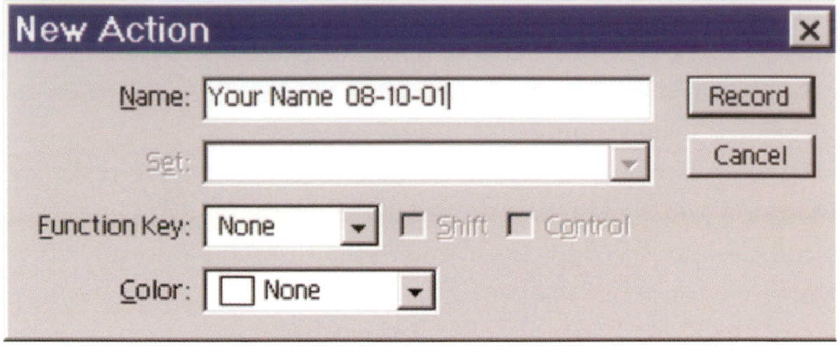

It is recommended that the action name includes who is working on the image and the date. Click on the record button. Starting a new action automatically turns on the recorder (record button at the bottom will be highlighted as in Figure 6.93.) The action palette will show the name of the new action and any adjustments to the image will be recorded from this point on.

Most commands can be accomplished from the buttons at the bottom of the palette without using the drop-down menu. They include from the left:

- Stop Recording
- Begin Recording
- Play Current Selection
- Create New Set
- Delete Selection
- Create New Action

Figure 6.93

Now let's build a sizing action. (Beware – anything you do now is being recorded.)

Choose the Image > Image Size command. Notice how large the image is and the resolution in Figure 6.94. You will size the image for printing.

Turn Resample Image off – you do not want to add or remove pixels from the image. Enter a value of 300 in the Resolution box. Note how the image is resized, but the pixel dimensions remain the same (Figure 6.95).

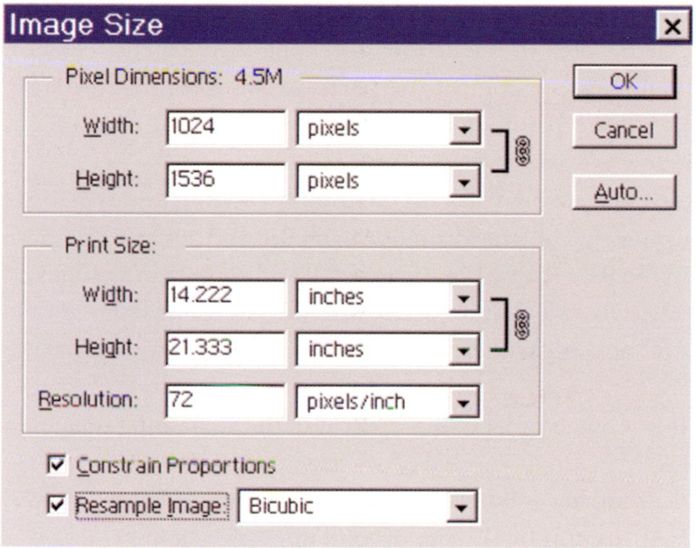

Figure 6.94

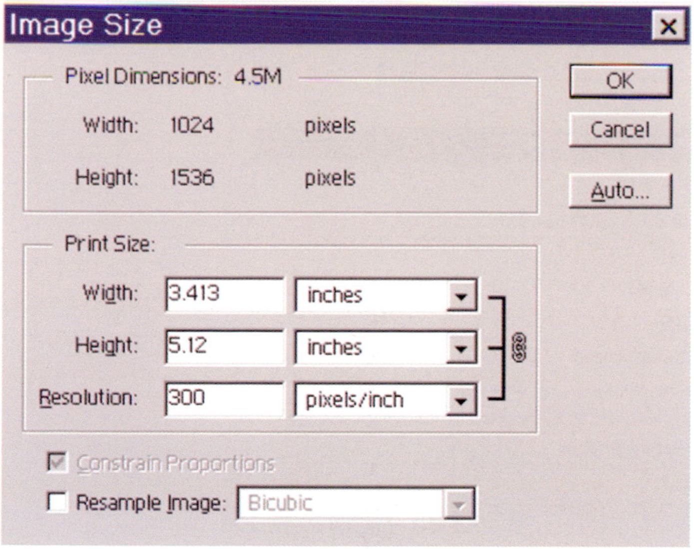

Figure 6.95

Click OK to apply the image size to the image.

Choose File > Save As > Your Folder > Sizing Folder, then click on the Save button.

Click on the "×" button in the title bar of the image to close it.

Click on the Stop Record button at the bottom of the actions palette.

The image has been resized and saved in another folder with the same file name. All the actions have been recorded in the sizing action so they can be used for additional images.

Now let's look at what was recorded in the action palette. To expand the actions, click on the arrow next to the action. To expand the palette, click on the "–" button in the top-right corner of the palette as shown in Figure 6.96. Note the image size and save parameters, showing exactly how the image was resized and how/where it was saved.

Now you need to save the action. Click on the topmost action, in this case Set 1, as shown in Figure 6.97. Use the drop-down menu and choose the Save Actions command. In the Save dialog box (Figure 6.98) choose Your Folder and name the action Sizing. Observe that it has an .atn file extension. The .atn file extension identifies all action palettes. Click on the Save button to save the action in Your Folder.

Now you will use it in a batch process and apply it to the remaining images in your sizing folder.

Choose the File > Automate > Batch command. The Batch dialog box will open (Figure 6.99). At the top of the dialog box appears whatever action is loaded in the action palette, in this case Your Name and the date.

Figure 6.96

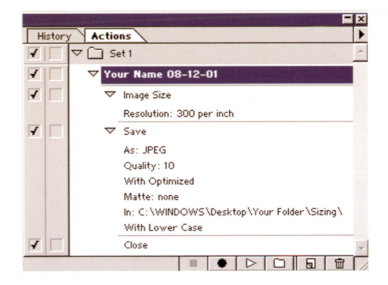

Figure 6.97

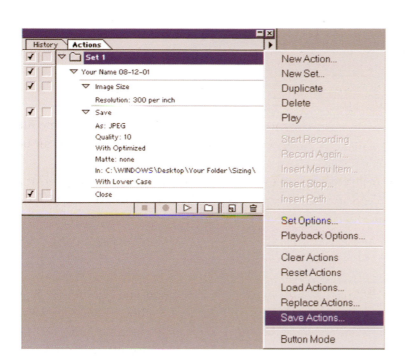

Figure 6.98

Click on the Choose button and choose the folder where the images are located that you want to resize. Leave the other settings in the dialog box at default.

Click OK to apply the process.

Each image is opened, resized, and saved to the new folder that you created, just as you did while recording the actions. I think you can see the value of the action, just in time savings; imagine if you had hundreds of images to resize or any other function that you wanted to apply to all images.

Figure 6.99

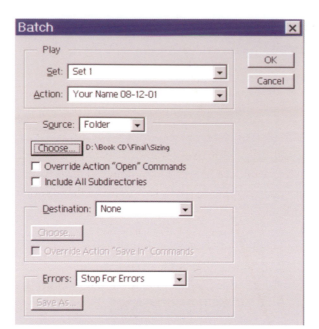

Figure 6.100

Let's explore the action palette a little more. Clear the actions from the palette. Choose File > Open > Your Folder > Sizing > Size_1.jpg. This will open one of the images that you resized.

Start a new action with your name and date as shown in Figure 6.100. Note the black border around the image; you will crop this out of the image. Select the crop tool, turn off the fixed target size in the options palette and place a marquee around the image as shown in Figure 6.100. Press the Enter key to apply the crop. Observe what appeared in the action palette as shown in Figure 6.101. Now apply the basic enhancements to the image as discussed earlier:

- Levels command
- Unsharp mask

Watch each action being recorded in the palette. Click on the arrow next to each action to open it – each action is described in detail – showing the exact values applied to the image. When you have finished enhancing the image click the Stop Record button at the bottom of the palette. The action palette should look somewhat like Figure 6.101.

Save the action: remember, select the uppermost action.

Open the drop-down menu and choose the Save Actions command. As a rule all image action files should be saved in the same folder as the processed image, and with the same name. There will be two files with the same name – one will have a graphics extension such as .tif or .jpg, the other will have an .atn extension.

Close the image. Do not save.

Notice the two columns to the left of each action in Figure 6.102. The column with the check marks allows you to turn off that individual action. The

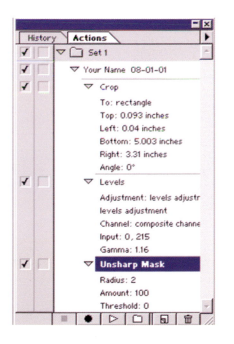

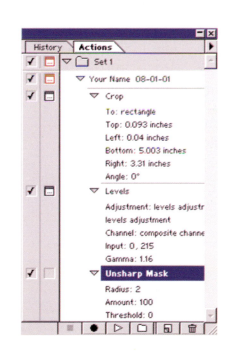

Figure 6.101 (left)

Figure 6.102 (right)

column with the rectangle pauses the application at that action so you can make individual adjustments. Click in the box next to crop and levels, so the application will pause at these actions. You will pause on these two actions, because they will probably be different for each image.

Choose the File > Automate > Batch command. Click on Choose > Your Folder > Sizing to locate the images you want to apply the actions to. Click OK.

The application will pause on the crop and levels actions allowing you to make individual adjustment. As you click OK the play actions will continue, until you have done each image in the folder.

Observe that each image is open on the desktop now, because you didn't record saving the images. Save and close each image.

Note: If more work needs to be done on the image a new action can be created at any time. The new action's name should follow the same parameters as the original – your name and the date. There is no limit to the number of actions that can be added to the action palette.

A GUIDE TO WHAT ACTIONS APPEAR IN THE PALETTE

From the menus

- File: Only the commands within the top two menu sections
- Edit :   All commands with the exception of Undo
- Image: All commands
- Layer: All commands
- Select: All commands
- Filter: All filters

From the toolbar

Records with parameters:

- Crop
- Marquee
- Magic wand
- Move
- Text
- Gradient
- Fill
- Eyedropper

Records with no parameters:
- Airbrush
- Paintbrush

- Rubber stamp
- Pencil
- Blur/sharpen
- Dodge/burn/sponge
- Eraser

Does not record:

- Pen tool
- Ruler
- Hand
- Zoom

LAYERS

When you create, import, or scan an image in Photoshop, the image consists of a single layer. You can add layers to the image, allowing you to organize your work into distinct levels.

You can draw, edit, paste, and reposition elements on one layer without disturbing the others. Until you combine, or merge, the layers in the original image, each layer remains independent of the others. This means you can experiment freely with various graphics, type, opacities, and blending modes. In addition, special features such as layer masks, layer effects, and adjustment layers let you experiment with and apply sophisticated effects to the layers in your image.

In Photoshop, the first layer of an image is called the background layer, analogous to the base layer of a painting. You cannot change the stacking order of a background layer or apply a blending mode or opacity (unless you convert it to a normal layer).

To avoid working on the original image most work in Photoshop should be done in layers.

CREATING LAYERS

Adobe Photoshop lets you create up to 100 layers in an image, each with its own blending mode and opacity. However, the amount of memory in your system may restrict the number of layers possible in a single image to less than this.

Note: Images created using the Transparent option in the New dialog box are created without a background. Images with no background, as well as images with layers, can be saved only in the Photoshop format. Newly added layers appear above the selected layer in the layers palette. You can add layers to an image in a variety of ways:

- By creating new layers or turning selections into layers.
- By converting a background to a regular layer or adding a background to an image.
- By pasting selections into the image.
- By creating type using the horizontal type tool or vertical type tool.

Ensure that the layers palette is open on the desktop.

Choose File > Open > Chk 169.jpg from the CD. Drag the layers palette under the image as shown in Figure 6.103.

Figure 6.103

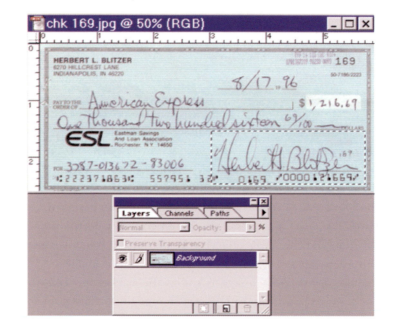

Figure 6.104

Select the signature block, including the check number beneath it with the rectangle marquee tool.

Choose Layer > New > Layer Via Copy. Observe the layers palette – you now have a new layer in the palette called Layer 1. The new layer is created, containing only the pixels contained in the area you selected. Note that the marquee is removed from the image and you can see no difference in the image. But there is a second signature block directly on top of the original. Select the move tool from the toolbox and move the new layer around the image. Place it back over the original signature block.

All layers should be named as they are created so they can be located easily for future use. Double-click Layer 1 in the layers palette. The layer option dialog box will open (Figure 6.104). Name the layer Signature. Click OK.

Build another layer using the same procedure. Click on the background layer to make it the active layer before you begin.

Select the numerical amount block with the rectangle tool.

Choose Layer > New > Layer Via Copy. Name the layer Amount.

The layers palette should appear as shown in Figure 6.105. There should be a signature layer, an amount layer, and the background. Observe the two layer icons in the palette; both have a checkered background, indicating that it is transparent. All layers by default are created on a transparent background.

You're probably wondering by now why you would even create layers such as these. What's the purpose? These could be used for many things, to apply enhancements to individual layers or possibly move the layers to a new image to build a composite. This we will accomplish later.

Observe the two columns to the left of the layers. The first column with the eye allows you to turn off the layer for viewing; the second column indicates the active layer.

Figure 6.105

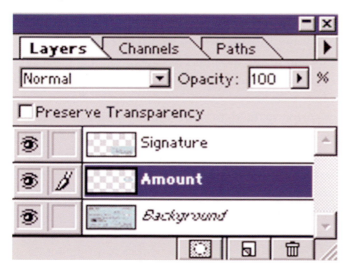

Figure 6.106

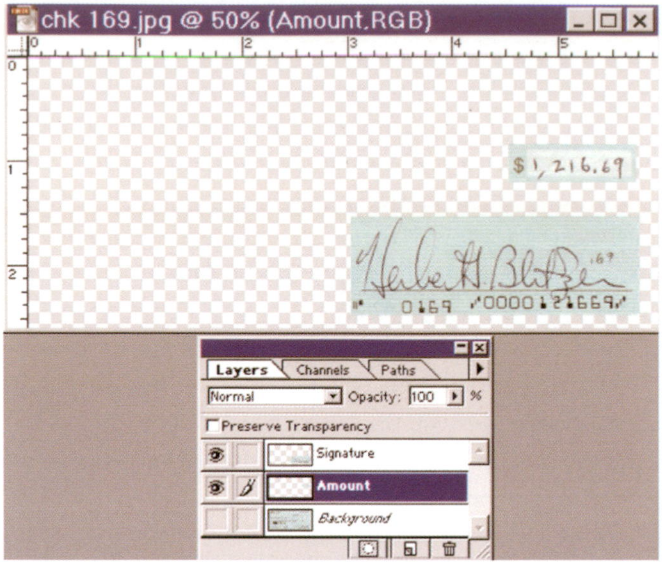

Click on the eye icon to the left of the background. This hides the background and the signature and amount is all that appears in the image as shown in Figure 6.106.

Click on the space where the eye was and the background reappears.

Click on the eye icon to the left of the background layer. Only the new layer you created remains on a transparent background. Click on the space where the eye was to the left of the Background layer. The background reappears.

OPACITY

The opacity of each layer can be controlled individually. Click on the signature layer to make it active. Select the move tool and drag the signature layer in the image so it partially overlaps the amount layer as shown in Figure 6.107. The signature layer appears on top of the amount layer, because of the stacking order in the layers palette. Layers are stacked from the bottom of the palette to the top.

Click on the drop-down arrow to the right of the opacity setting in the layers palette. A slider appears for adjusting the opacity. The opacity can also be adjusted from the keyboard number pad or entering a value in the window. Adjust the slider to set the opacity at 40% as shown in Figure 6.108. Observe that the amount layer appears through the signature layer, because the opacity of the signature layer has been reduced. Return the opacity to 100%.

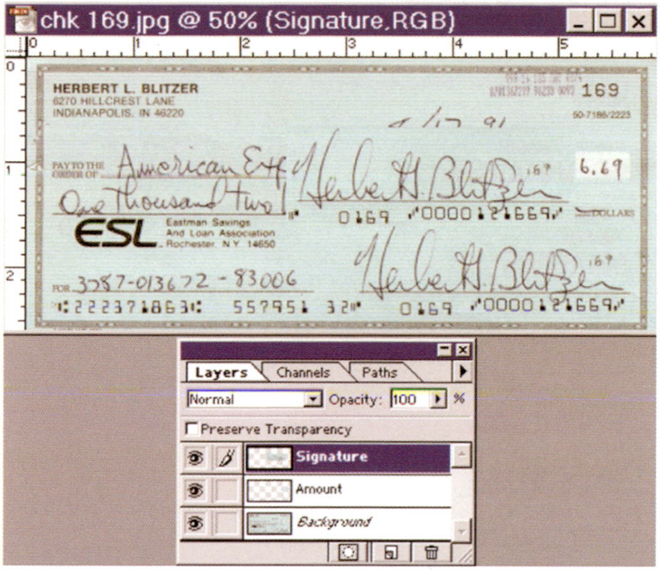

Figure 6.107

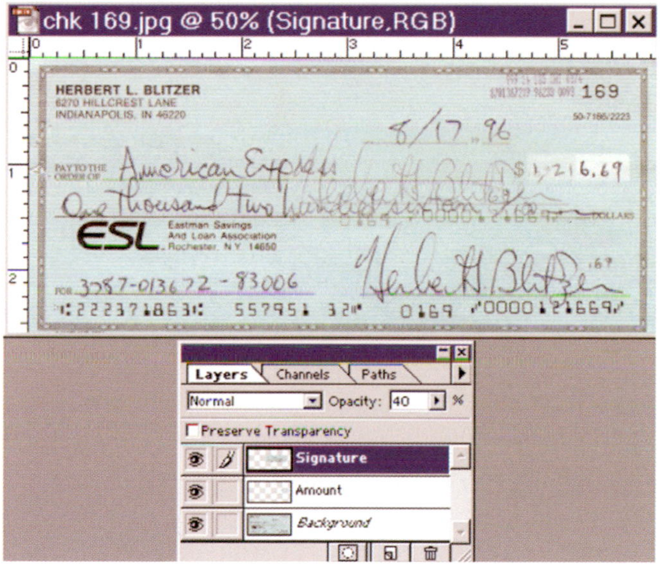

Figure 6.108

STACKING ORDER

The stacking order determines whether a layer appears in front of or behind other layers. It makes no difference in what order you create layers; they can be placed in the correct order after the image is finished. In the layers palette, select the signature layer to make it active. In the layers palette drag the layer down under the amount layer. When the highlighted line appears in the desired position, release the mouse button (Figure 6.109).

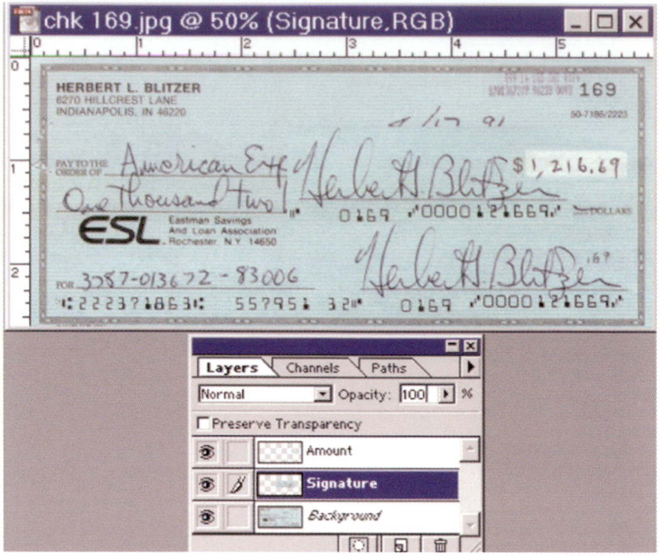

Figure 6.109

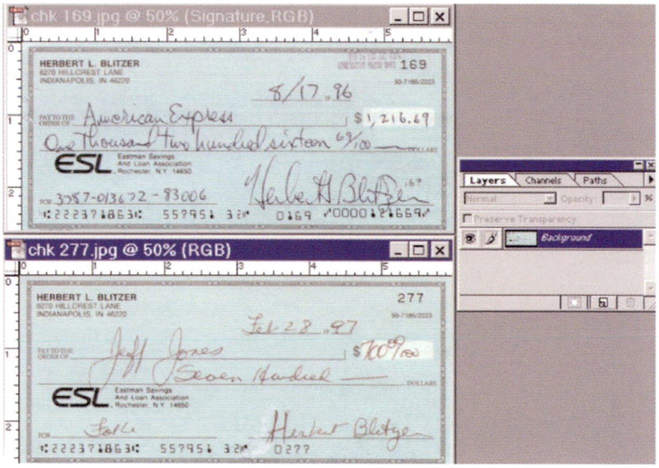

Figure 6.110

Note: By default, the background cannot be moved from the bottom of the layer list. To move the background, it must be first converted to a layer.

Note now that the amount appears on top of the signature in the image.

The stacking order is in direct relationship to how the layers appear in the layers palette. Select the move tool and return the signature copy so that it overlays the original.

Choose File > Open > Chk 277.jpg from the CD. Arrange the checks and the layer palette as shown in Figure 6.110. Observe that Chk 277 only has a background; no layers have been added.

Make Chk 169 the active image.

COPY LAYERS BETWEEN IMAGES

Click on the signature layer you created in the layers palette to make it active. Place the mouse on the layer name in the layers palette and drag it onto the second image. Release the mouse button and a new layer is created. Use the move tool to align it over the original signature as shown in Figure 6.111. Observe that a new layer is automatically created in the layers palette.

Layers can also be copied between images by dragging the layer itself from one image to the other or using the Copy and Paste commands. Make Chk 169 the active image, click on the amount layer to make it active, and choose the move tool.

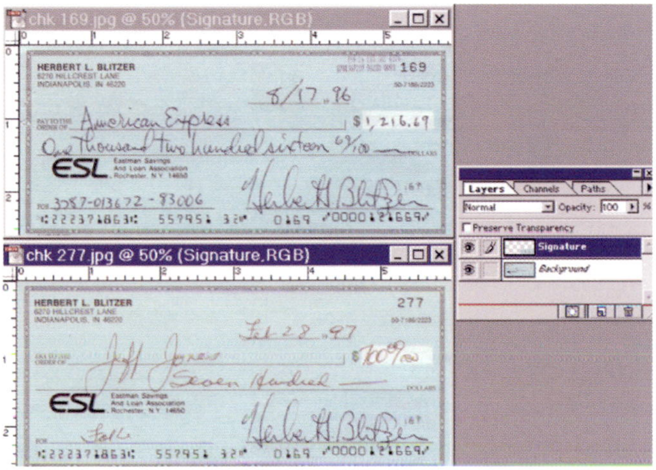

Figure 6.111

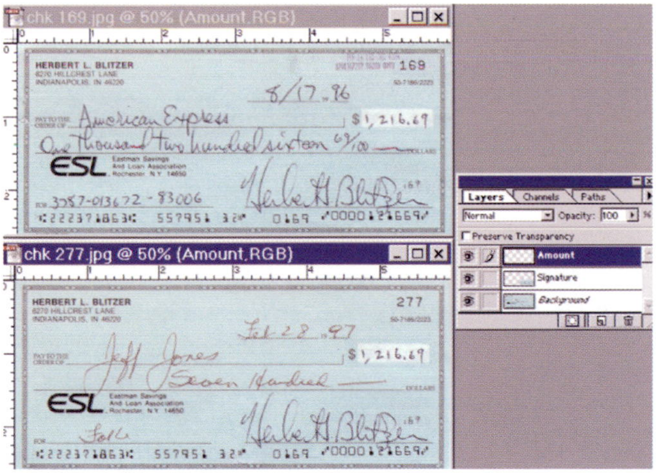

Figure 6.112

Figure 6.113

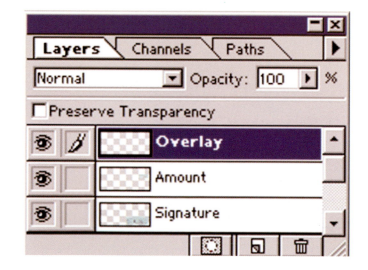

Click on the amount in the check and drag it down to Chk 277 as shown in Figure 6.112. Layers can also be transformed (sized, rotated, etc.), using the transform command. Ensure that the amount layer is active in Chk 277. Choose the Edit > Free Transform command. Observe that the transform box appears around the amount layer in the image (Figure 6.113). Size the layer to make it fit.

Remember to hold the Shift key to maintain proportion. Press the Enter key to complete the transformation.

CREATING A BLANK LAYER

Ensure that Chk 277 is the active image. Click on the new layer icon at the bottom of the layers palette. You can't see it in the image, because it is transparent, but a new layer is created in the layers palette. Double-click the new layer in the layers palette and in the layer options dialog box name the layer Overlay, then click OK. The layer palette should appear like Figure 6.114.

Figure 6.114

Select the line tool from the toolbox. Double-click the line tool to open the line tool option box.

Enter a value of 5 pixels in the weight box and place a check mark in the start arrowheads box. Set all other parameters as shown in Figure 6.115. Draw two arrows pointing to the two layers you created in the check as shown in Figure 6.116. The arrows are on the transparent overlay layer and do not affect the original image.

Figure 6.115

Figure 6.116

TYPE LAYERS

Layers are created automatically when you use the type tool. There are four type tools in the type toolbox. The type tool and vertical type tools let you create colored type that is stored in a new type layer. You can edit the text at any time using the type layer. The type mask and vertical type mask tools let you create selection borders in the shape of type. Type selections appear on the active layer, and can be moved, copied, filled, or stroked just like any other selection.

You enter text by selecting a type tool and clicking in the image to set an insertion point. The Type Tool dialog box lets you enter the text and specify formatting attributes. Once you have created type, you can edit the contents, attributes, and orientation at any time by using its type layer. Type layers save with the image. You can move, restack, copy, and change the layer options of a type layer just as for a regular layer. You can also apply these commands to a type layer and still be able to edit the text.

Select the type tool from the toolbox. Set the insertion point by clicking next to one of the arrows you created. The type dialog box will open. If necessary, drag it off the image so you can see the insertion point. Set the parameters as shown in Figure 6.117. Enter your initials in the text box as shown. The initials should also appear in the image where you placed the insertion point. Click OK to apply the type to the image.

Figure 6.117

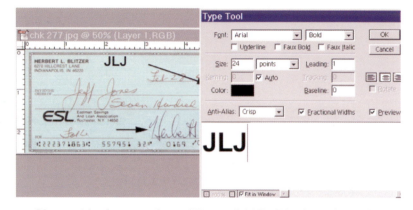

Observe the layers palette (Figure 6.118). The layer has a large "T" denoting type layer. Use the move tool to align the type next to the arrow.

Figure 6.118

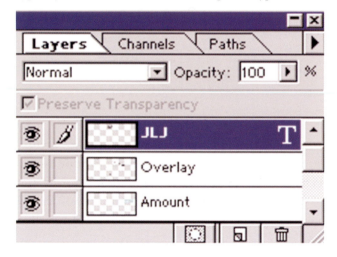

DUPLICATING LAYERS

Drag the Type layer you just created to the New Layer icon at the bottom of the layers palette. Release the mouse button when the cursor darkens the icon. You have created a type copy.

The layers palette should appear as shown in Figure 6.119. There are two type

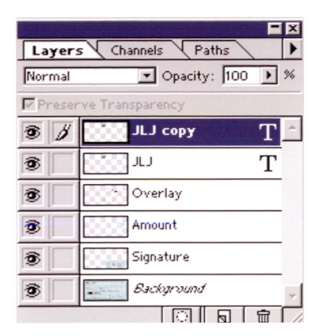

Figure 6.119

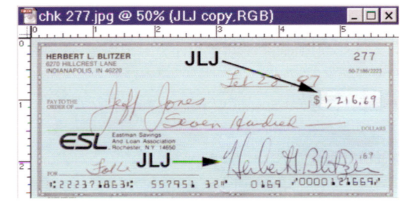

Figure 6.120

layers with your initials now. Use the move tool to drag the new layer off the top of the other, placing the initials by the second arrow you created. Your image should appear as shown in Figure 6.120. Both the initials should be aligned with the arrows in the image. To edit type layers, double-click on the layer in the layers palette. The dialog box will open and you will have full control over the type.

DELETING LAYERS

One of the most powerful parts of layers is if you make a mistake it does not affect your original image. It is on a layer and any layer can be removed by dragging it to the trashcan at the bottom of the layers palette.

MERGING LAYERS

Layers use a lot of memory. You will not notice it on this image, because it is a very small file size. But when you start working on large high-resolution images, memory will be consumed rapidly. One way of reducing this memory usage is to merge the layers together as you finish. One note of caution: do not merge layers until you have finished editing them. We will start with merging type layers, because they must be handled a little differently. Ensure that the type layer copy is active. Click on the triangle at the top right of the layers palette to open the drop-down menu. Observe that the merge options are all dimmed (not available), because type layers must first be converted to standard layers before they can be merged.

Choose the Layer > Type > Render Layer command. Note what happened in the layers palette (Figure 6.121). The layer is no longer a type layer; it is a standard layer like all the rest. Double-click the copy layer in the layers palette; you can no longer edit it as type.

Make the other type layer active and repeat the steps above to render it also. The layers palette should appear as shown in Figure 6.122. Both type layers are standard layers; they no longer have the "T" indicating type layer. Click on the copy layer to make it active. Click on the triangle at the top right of the layers palette to open the drop-down menu. Observe now that you have merge options available. Click on the Merge Down command and observe the layers palette. The initials layers merged together using the name of the lower layer. They are now one layer. Select the move tool from the toolbox. Move the initials in the

Figure 6.121

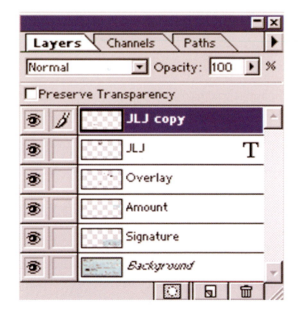

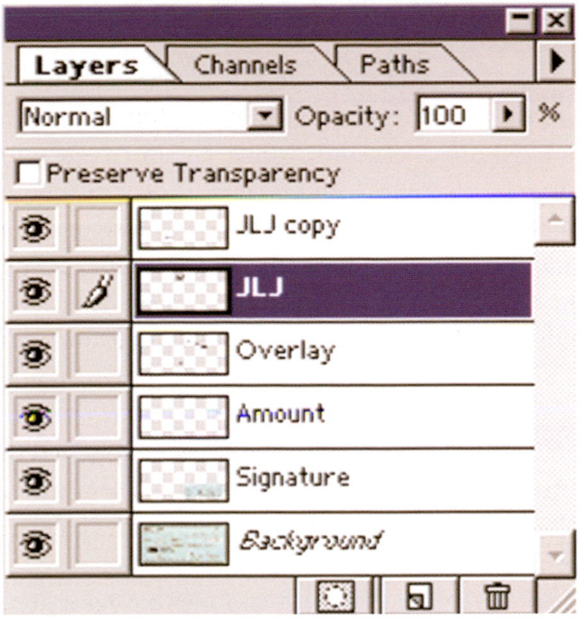

Figure 6.122

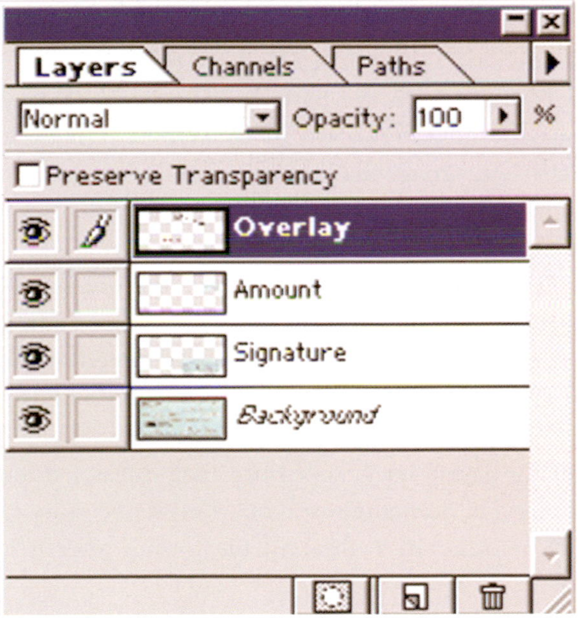

Figure 6.123

image; they now move as one, because they are on the same layer.

Repeat the above procedure; merge the initials layer with the overlay layer. Now the initials and arrows move as one object. The layers palette should appear as shown in Figure 6.123. The type and arrows are now on the overlay layer. Use the move tool to check this out. They should all move as one object.

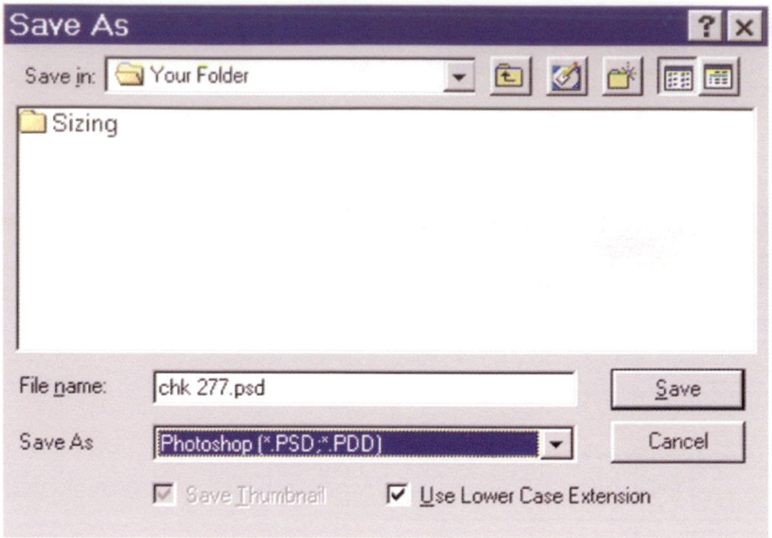

Figure 6.124

FLATTENING ALL LAYERS

When you are done editing, you can reduce the image back to its background layer by flattening the image. This is necessary for saving the image in any other format than Photoshop (PSD), such as TIFF, JPG, or EPS. Let's test this out.

Choose the File > Save As command. In the Save As dialog box (Figure 6.124), click on the drop-down arrow next to the Save As box. Observe that you have only one option, ".PSD,PDD." Any image that has layers cannot be saved in any other format without flattening the image. Leave the image with the same name and save it as a PSD format in your folder.

Now let's flatten the image. There are a number of ways of doing this; we will use the layers palette. Make certain all of the layers are visible. Choose Flatten Image from the layers palette drop-down menu. The image is reduced to the background layer as shown in Figure 6.125. Now the image can be saved as a TIFF, JPEG, etc.

Realize that you cannot get the layers back, so ensure that you are done editing the image prior to flattening it. Sometimes you may find it necessary to save two copies, one as a PSD, another in a different file format after it is flattened.

Close all images. Return the palettes to their default position.

ADJUSTMENT LAYERS

Another type of layer is the adjustment layer. Adjustment layers can be used on the entire image or on selected areas of the image. Like any other layer it can be discarded if it is not what you want. If you remember earlier when you did the

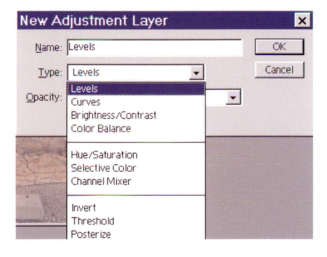

Figure 6.125

Figure 6.126

levels on an image it applied directly to the image. Applying levels as an adjust-ment layer gives you the option of removing it at a later time.

Choose File > Open > Image_1.jpg from the CD. Move the layers palette next to the image. Observe that there are no layers, only the background appears in the palette. Choose Layer > New > Adjustment Layer. The New Adjustment Layer dialog box will open. Click on the Type drop-down arrow and choose Levels from the list shown in Figure 6.126. Click OK.

The levels dialog box will open: make the adjustments as you did before to the image. Click OK and observe the layers palette in Figure 6.127. A new levels

Figure 6.127

layer has been created. The contrast icon on the right indicates it is an adjustment layer.

Click the eye icon to the left of the levels layer. It hides the adjustments and you can see what the original image looked like.

Double-click on the levels layer in the palette. The levels dialog box opens again for additional adjustments.

The adjustments can also be applied to other images by dragging the layer onto the image. Observe the list in Figure 6.126. Other adjustment layers can also be applied to the image.

Experiment with the adjustment layers and then flatten the image to the background.

CHANNELS

Click on the Channels palette tab in the Layers palette group.

Every Adobe Photoshop image has one or more channels, each one storing information about color elements in the image. The number of default color channels in an image depends on its color mode. For example, a CMYK image has at least four channels, one each for cyan, magenta, yellow, and black information. Think of a channel as analogous to a plate in the printing process, with a separate plate applying each layer of color. In addition to these default color channels, extra channels, called alpha channels, can be added to an image for storing and editing selections as masks.

An image can have up to 24 channels. By default, Bitmap-mode, grayscale, duotone, and indexed-color images have one channel; RGB images have three; and CMYK images have four. You can add channels to all image types except Bitmap-mode images.

Click on each of the channels: Red, Green, and Blue. Each of the channels appears as shades of gray – the darkest shade in the channel is the darkest Red, Green, or Blue, respectively.

CHANNEL MIXER

The Channel Mixer command lets you modify a color channel using a mix of the current color channels. With this command, you can do the following:

- Make creative color adjustments not easily done with the other color adjustment tools.
- Create high-quality grayscale images by choosing the percentage contribution from each color channel.
- Create high-quality sepia-tone or other tinted images.

Select the RGB composite channel in the channels palette.

Choose the Image > Adjust > Channel Mixer command. The Channel Mixer dialog box will open. Apply the setting as shown in Figure 6.128.

What you have done is eliminated all the tones but red and enhanced the red tones so there is more detail. Click OK to apply the changes. This can be very useful in color images, when trying to enhance a particular color.

There are many uses for the channels, too numerous to cover them all here, for instance making selections and masks. Experiment with them a little and you'll discover how powerful they are.

Close the image, do not save.

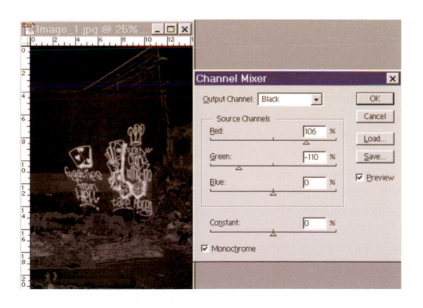

Figure 6.128

FILTERS/PLUG-INS

There are many filters/plug-ins included with Photoshop; some work quite well for forensic work, others are designed strictly for the artistic side of the house.

Adobe Photoshop also supports plug-in filters developed by non-Adobe software developers. Once installed, plug-in filters appear at the bottom of the Filter menu and work the same as built-in filters. If you are interested in creating Photoshop-compatible plug-in modules, contact Adobe Systems Developer Support.

Plug-in modules are software programs developed by Adobe Systems, and by other software developers in conjunction with Adobe Systems, to add features to Adobe Photoshop. A number of importing, exporting, automation, and special effects plug-ins come with your program – automatically installed in the Plug-Ins folder. If you want to use a plug-in that is installed in another folder, you must change the preference settings so that Photoshop uses the other plug-ins location. Once installed, plug-in modules appear as options in the Import, Export, or Automate menus; as file formats in the Open, Save As, and Save a Copy dialog boxes; or as filters in the Filter menus.

Note: If many plug-in modules are installed, Photoshop may not be able to list all the plug-ins in their appropriate menus. Newly installed plug-ins will appear in the Filter > Other submenu.

IFI has designed plug-in filters for Adobe Photoshop that will help in sizing images, removing backgrounds, and analyzing images, and a grid to help align objects in the image. Once installed, all the filters will be located under Filter > IFI.

FFT (FAST FOURIER TRANSFORM) FILTER

The example that will be used is a photograph of a fingerprint on a picture in a newspaper. The FFT filter will be used to subdue the dot pattern appearing in the background.

Note: This filter can help subdue many backgrounds with a consistent pattern running horizontally, vertically or diagonally. It works to separate high-frequency from low-frequency information. The dot pattern is evenly spaced horizontally and vertically making it high-frequency information. Conversely, the fingerprint is somewhat random in its spacing, making it low-frequency information.

Choose File > Open > Paper_Print.jpg from the CD.

Duplicate the image prior to beginning – the filter applies the changes to the original image. We will apply the filter to the Paper_Print copy.

Choose the Filter > IFI > Forward FT command. The filter dialog box will

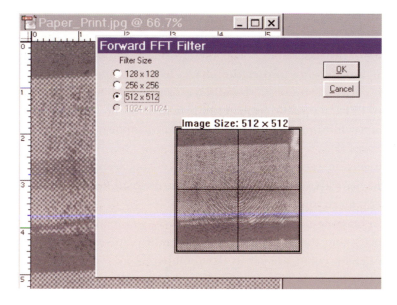

Figure 6.129

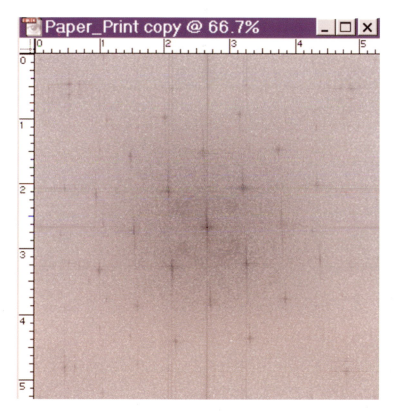

Figure 6.130

appear (Figure 6.129) allowing you to select the filter size and area of the image to be included. Select the filter size that will encompass the majority of the fingerprint. Hold down the mouse button on the crosshairs in the center of the box and drag the box to center it over the image. In this example the fingerprint takes up the entire image area and would require the 512×512 filter. Click OK – the Forward FFT filter will be applied to the image.

The low-frequency information on the fingerprint will appear in the center as a dark shaded area (Figure 6.130).

Outside the fingerprint area will be some dark spikes appearing like a pattern and possibly some vertical or horizontal lines. This is the high-frequency information or the dot pattern you are trying to eliminate.

You must use a masking technique and fill the areas you want to keep (fingerprint) with black and the areas you want to remove (spikes) with white. Ensure that the default colors are set in the toolbox (black foreground and white background). Select the ellipse selection tool from the toolbox. Hold down Shift + Alt to maintain a circle from the center. Place the cursor on the very center of the fingerprint area and drag a marquee similar to the one in Figure 6.131.

Figure 6.131

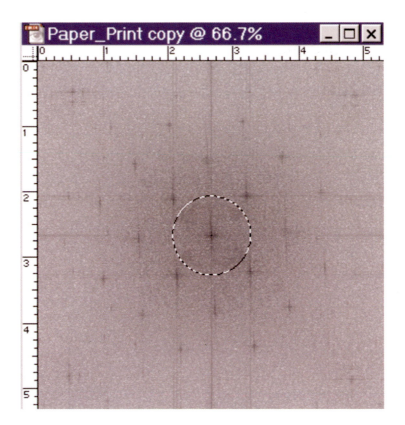

Choose the Edit > Fill command or the paint bucket to fill the area with the foreground color (black). Ensure that the opacity is set to 100% in the Fill dialog box.

Choose the Select > Inverse command to invert the selection, to select everything but the black area.

Select the Edit > Fill command or the paint bucket to fill the area with the background color (white). This procedure masks the image – black is what you want to keep – white is what you want to remove.

Note: Ensure that you fill the areas with 100% black and white. None of the areas of the original image can show through the fill color. Press Ctrl + D to remove the selection from the image. It should appear similar to Figure 6.132.

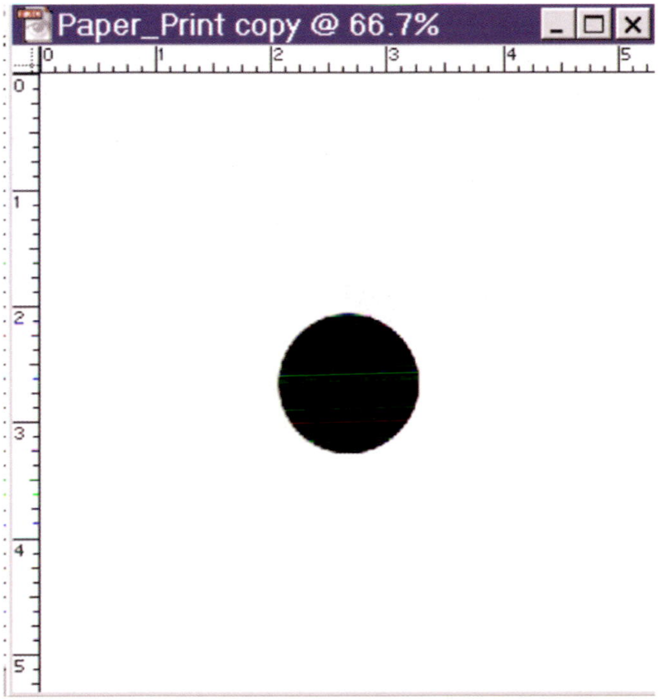

Figure 6.132

MASK INVERSE FILTER

Select the Filter > IFI > Mask Inv command. This will redraw the image, applying the FFT filter, subduing the background dots and making the fingerprint more readable. The image will appear as displayed in the illustration. At this point the FFT is finished.

Any further adjustments to the image can be made individually, such as contrast, unsharp mask, density, etc.

Note: Remember this is a duplicate image; the original is still intact in its original file.

Figure 6.133

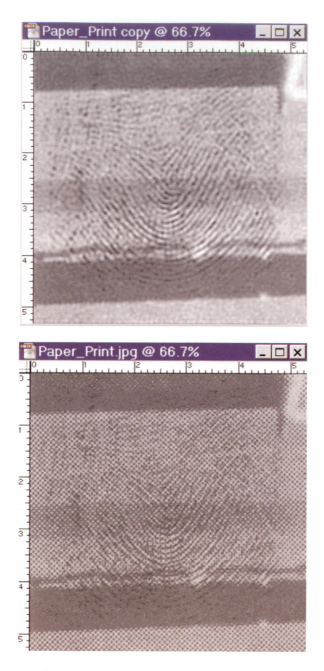

Figure 6.134

The filter is applied in Figure 6.133; the dots in the background are subdued, making it easier to read the fingerprint. Figure 6.134 is the original; notice the difference.

Make the Paper_Print copy the active image. Choose File > Save As > Your Folder. Leave the image name as Paper_Print Copy and save it.

Close all images.

SIZING FILTER

We all know that we can size images to 1:1 as we photograph them, but sometimes it is easier to just place a scale in the image and size them after the image is in Photoshop. This can be accomplished by using the sizing filter supplied by IFI.

Note: Scanned images default to 1:1 when scanned.

There must be a ruler in the image or a known measurement of some kind in the image to set the standard for the filter.

Choose File > Paper_Print2 from the CD.

Choose the View > Show Rulers command if they are not already showing.

Choose the Filter > IFI > Sizing command. The sizing dialog box will open as shown in Figure 6.135.

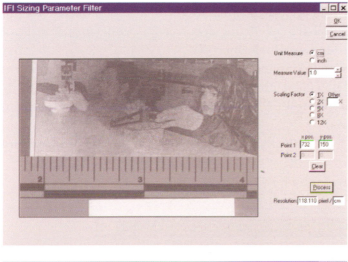

Figure 6.135

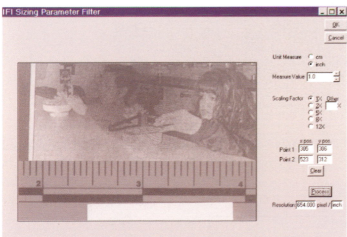

Figure 6.136

Set the unit of measure to inches to match the scale that is in the image. Enter the value you will be selecting from the ruler, in this case 1 inch. Click on the Clear button to clear all other data in the box.

Taking care to be accurate, click a beginning point on the ruler – in this case the 2 inch mark. Click an ending point on the ruler – the 3 inch mark. You have now set 1 inch in the filter. As you clicked each point, the *x* and *y* coordinates were set in the boxes.

Click the Process button. The value should be similar to those in Figure 6.136. If not, press the clear button and try again.

Note: The sizing filter does not automatically resize the image to 1:1 – this must be done manually by entering a new resolution.

Record the resolution value; it should be very close to 651. This value will be entered manually in the Image Size dialog box of the original image. Click the OK button to close the dialog box.

Select the Image > Image Size command. The Image Size dialog box will open. Ensure that the measurements are set correctly – centimeters or inches – to match the measuring device in the image.

Turn off the Resample Image control. Enter the new resolution. Observe how the image resizes in the width and height boxes.

Click OK. The image is resized 1:1. The image did not change on the screen, but observe the rulers.

You can double-check the sizing by using the measure tool in the toolbox and the info palette.

Choose File > Save As > Your Folder > Save.

Figure 6.137

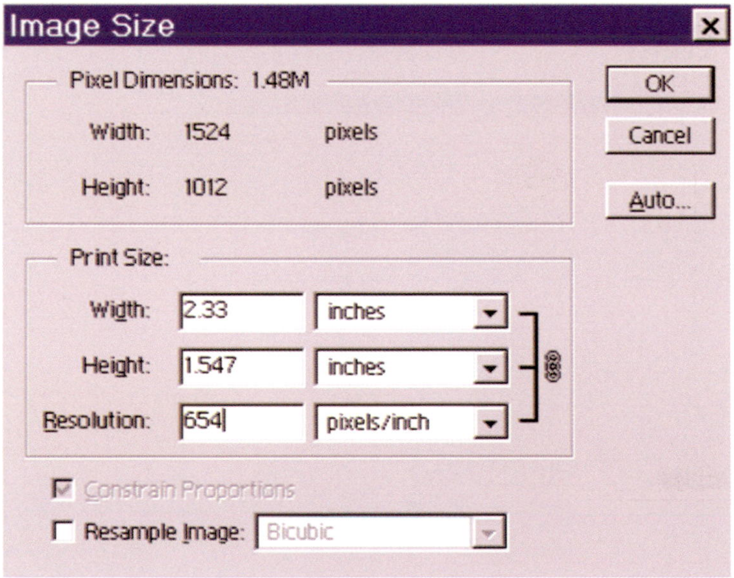

CHAPTER 7

IMAGE APPLICATIONS

Before we begin the laboratory section we should discuss a few lighting methods for close-ups of evidence that work for both digital and film. These are common techniques and if used properly will render great detail in your photographs.

LIGHTING

The following lighting methods are effective for photographing various evidence subjects. The effects should be previewed in order to select the best lighting technique for the evidence subject.

DIRECT LIGHTING

Direct lighting (Figure 7.1) uses normal copy lighting with one or more light sources at 45° angles.

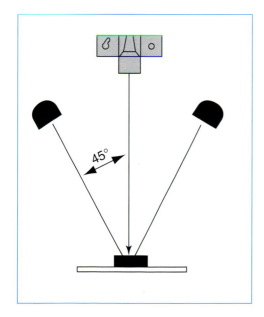

Figure 7.1

DIRECT REFLECTIVE LIGHTING

Light is reflected directly off the subject into the lens. Place the subject at a 10° angle from the lens to film plane and place the light source at 10° angle from the subject (Figure 7.2). The light source reflects at a 20° angle into the lens. The light source may need to be diffused to prevent hot spots. This method creates very high contrast.

Figure 7.2 (left)

Figure 7.3 (right)

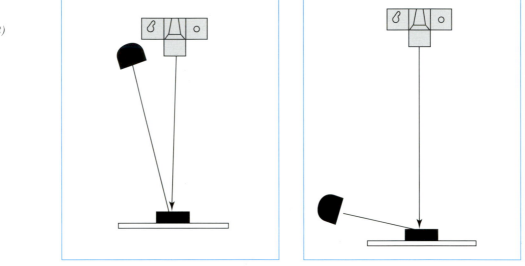

OBLIQUE LIGHTING

Oblique lighting (Figure 7.3) uses a light source at a low angle, usually to show detail by creating shadows in the subject surface. It is commonly used when photographing impressions, tool marks and certain types of fingerprints.

BOUNCE LIGHTING

Light is bounced off a white or reflective surface (Figure 7.4). The bounce surface may be positioned at different locations (above or to one side of the subject) to create the desired effect. This usually produces even, non-glare lighting with low contrast.

DIFFUSED LIGHTING

An opaque material is placed between the light source and the subject to diffuse the light (Figure 7.5). This usually results in even lighting with reduced reflections and hot spots.

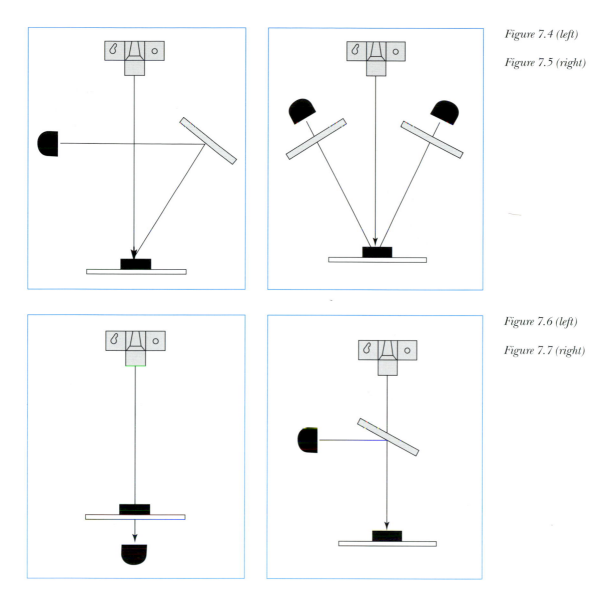

Figure 7.4 (left)

Figure 7.5 (right)

Figure 7.6 (left)

Figure 7.7 (right)

TRANSMITTED LIGHTING

With transparent subjects the light source is transmitted through the subject toward the lens (Figure 7.6). The angle of the transmitted lighting is adjusted from 90° to 45° for the desired effect.

FRONT DIRECTIONAL OR AXIS LIGHTING

A clear piece of glass is placed between the subject and lens at a 45° angle (Figure 7.7). The light source is positioned parallel to the film and 45° to the glass. While the light is transmitted through the glass, some is reflected downward directly onto the subject. This technique is effective when pho-tographing fingerprints on mirrors and into glasses or cups.

LATENT FINGERPRINT

Once photographers understand digital imaging they will find the digital camera to be an excellent imaging tool. A digital camera gives you the ability to see the captured image immediately. It also gives you the ability to make a judgment on exposure and quality immediately. The image can also be trans-mitted electronically for others to evaluate.

FINGERPRINT CAPTURE

Many times it is necessary to photograph fingerprints that are still on the original surface. This may happen when a powdered fingerprint lies on a rough surface, and lifting the print onto tape would carry with it distracting back-ground information, or from a hard-to-reach area where there is a likelihood of getting a poor-quality lift from the original. It may have to be done in the field, or it may be possible to bring the evidence back to the lab. The techniques remain the same; it is just that you usually have more equipment to work with in the lab.

Fingerprints lifted using fingerprint tape can be photographed or scanned using a flatbed scanner. The preferred method is to photograph the lifted fin-gerprints with a digital camera. A typical scanner usually has about a 600-pixel resolution, versus 1500-pixel resolution or higher with a good digital camera. Higher resolution allows you to see much more detail in the fingerprint, which is essential for fingerprint comparison and enhancement of a poor print.

A digital camera adds another dimension to the forensic process by bringing a fingerprint image directly into the computer. Adjustments can be made instantly for lighting and exposure, ensuring the best possible capture of the fingerprint image. Once the image is digitized it can be enhanced, archived and stored in a database, and used to make on-screen comparisons.

Let's look at some specific types of fingerprint subjects and the best way to light and photograph them.

- Normal dusted prints can usually be photographed with no problem using 45° lighting.
- Fingerprint impressions left in soft substances such as wax, putty, clay, adhesive tape, and grease, or in dust, normally will need a cross-lighting technique. You can preview the results by using a handheld light source, such as a flashlight.

- Photographing fingerprints on a porous surface is a lot like the impressions, but requires a much lower angle for the lighting; usually it will be close to 90°.
- Glass and mirrors create special problems. A good method for glass is to place white card or cloth behind the glass, and use a low grazing angle of light. Mirrors are also going to require a low-angle or bounced technique for the lighting. Beware of reflections and focus carefully when working with mirrors.
- Perspiration prints on glass require somewhat of a special technique. Use back (transmitted) lighting and diffusion screen to highlight the fingerprint and soften the lighting.
- Ninhydrin fingerprints can be a problem with the digital camera. They are best photographed in black and white with a green filter. Many of the digital cameras have the feature where you can set the camera on a monochrome setting. This would work quite well for ninhydrin prints.

ALTERNATIVE LIGHT SOURCE

When a fingerprint is left on a surface, but traditional fingerprint powders cannot provide optimum results, the photographer may need to use fluorescing dyes to stain the print. Staining causes the fingerprint to glow when illuminated with a fluorescent light source.

Special techniques are required, because exact exposure, filtration, and even illumination are a must. Again, instant viewing of the captured image is of high value when using fluorescence and reflective applications. This is why a professional digital camera saves considerable time and expense associated with having to develop film, and possibly reshoot.

The colors that are visible to the eye represent only a small portion of the light spectrum, also known as the electromagnetic spectrum. Visible light, or white light, is a combination of all the visible colors. A beam of white light can be separated into the visible spectrum using a prism. The band of colors comprises violet, blue, blue-green, green, yellow, orange, red, and deep red. Each color represents a different wavelength of light. These wavelengths increase in the direction from blue to red along the length of the spectrum. The visible region of the light spectrum ranges from 400 to 700 nanometers (nm) in wavelength.

The areas extending in either direction beyond the visible spectrum are the invisible regions of light. Below violet from 200 to 400 nm is the ultraviolet region. Although we cannot see this light, it is reactive with photographic materials. Therefore, it is possible to produce images that may only be observed using photography. Extending just beyond the visible region in the other direction from

700 nm and higher is the infrared region of light radiation. The range of infrared light close to the visible spectrum is also photographically reactive.

The illumination of the subject (Figure 7.8) should consist only of the radiations needed to excite the fluorescence. All ambient illumination (room lights) must be excluded from the subject. Most often this is done in a darkened room or light-tight enclosure.

Figure 7.8

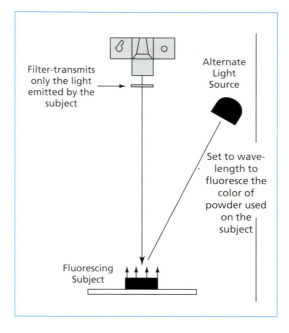

Use a light source that allows adjustment of the wavelengths of light or a filter on the light source that transmits the wavelength needed to fluoresce the powder and blocks all other light.

Use a filter on the camera lens to block all light other than that being emitted by the fluorescing subject.

The light will strike the subject and excite fluorescence. The fluorescence will then be recorded on the film.

We need to learn other photographic techniques to capture these images on a digital camera. Know the limitations of the camera. Due to the low light levels, a high ISO is required, which would eliminate the use of some of the inexpensive cameras. Maintain a constant distance of the light source to the object for consecutive exposures and establish a starting point by a camera meter reading or handheld light meter, then bracket the exposure by at least two stops. Using these simple rules you will capture quality images suitable for comparisons. Of course, you should now see the advantage of the digital camera, instant viewing of the captured image eliminates the need for bracketing.

One final note – practice, practice, practice!

QUESTIONED DOCUMENTS

Although most of the time accurate decipherment can be achieved by a visual study of the document, photography can be used to record what was originally written. Photography can, however, in itself be an essential tool for achieving many decipherments. It is probably one of the most effective tools at the document examiner's disposal. Photography accurately preserves the condition of the document and can help to show that there has been an erasure. Finally, it provides the means for demonstrating to others, especially judges and a jury, all of these findings.

Many photographic techniques are available. A few involve the use of special materials, but most do not. What is generally required is stricter control of various standard photographic steps rather than unusual techniques. A standard document photograph, that is, one in which sufficient care is taken in its preparation to record maximum detail and eliminate distortion, may of itself show what the original writing was.

Normally, however, there is a need at least to increase the contrast slightly in the final print over that of the original document. Weak areas may be intensified sufficiently in order to permit an accurate decipherment. Black-and-white photographs are the standard. But the most important consideration in any photograph intended to decipher erased writing is maintaining maximum detail.

A normal-size photograph usually gives the best results, although there is some advantage to enlarging the image. Too great an enlargement introduces effects similar to visual examination under high magnification, spreading the weak fragments so that they become harder to interpret and may even weaken some finer details. The degree of enlargement requires careful selection and possibly some experimentation with the problem at hand.

OBLIQUE LIGHT PHOTOGRAPHS

One of the standard means of deciphering erased pencil writing photographically is the use of oblique or low-angle side lighting. Whenever writing impressions remain from the original text this technique should be employed.

The most important consideration in making side-light photographs is the angle at which the light strikes the paper. A light–paper angle of 10–30° or occasionally less than 10° produces the best results. The actual angle used will vary from problem to problem and depends on how flat the paper surface can be held, and its texture.

VERTICAL ILLUMINATION

Another technique for illumination is with a light source and camera mounted vertically above the document. The light source is a ring illuminator, which fits around the camera lens. This means that the light reflected from the document strikes the lens parallel to its axis.

LOW-INTENSITY LIGHT

A further special lighting technique of value in deciphering erased pencil writing involves the use of low-intensity illumination, especially with a daylight source. It can be combined with the various types of photography already discussed.

FILTER PHOTOGRAPHY

Certain photographic filters may improve the decipherment of partially erased pencil writing. If the erased writing was written with a colored pencil rather than black, selection of a filter of contrasting color can intensify very weak fragments. If the overwriting in the erased area is in color rather than black, it can be weakened or eliminated by choosing a filter of similar color. For example, with an erasure of red pencil a green or blue filter helps to intensify the weak traces. If by chance the revised writing were made with a blue pencil, the blue would weaken the overwriting at the same time.

POLAROID SCREENS

Experiments using Polaroid screens in deciphering erased writing indicates that their use is of limited value. The screen can be placed either over the camera lens or between the light source and the document.

A pair of Polaroid screens, one over the lens and one on the light source, is sometimes more desirable. Rotating of one screen to reduce highlights shows slight improvement, but as a rule, this type of photograph offers no practical improvement unless one is dealing with an erasure on a glossy writing surface.

IMPROVING CONTRAST

There are cases in which the best exposure does not record all the details with sufficient contrast to interpret them accurately. Some improvement in contrast is indicated. This is easily accomplished with a digital image using available image enhancement software. Care must be taken that detail is not lost by too much contrast.

SUPERIMPOSED NEGATIVES

The use of multiple images provides a further special technique. Two or more duplicate images are prepared. Each is made under identical conditions of size and exposure, with exposure controlled to produce a thin negative. The images are carefully superimposed in exact register and printed. The procedure is known to result in a final print containing somewhat more information than can be obtained by other means.

BACK OF DOCUMENT

With many pencil-written documents containing an erasure, both the original and the new overwriting leave impressions in the paper. If the back of the sheet is blank, and sometimes even with limited writing behind the erasure, a better decipherment may be obtained by photographing that side for study. The embossings are recorded and emphasized by carefully controlled, oblique light photography. Embossing tools in the image-editing program can also help enhance the original.

SUMMARY

Digital photography can be a virtually indispensable method for deciphering erased pencil writing. But it is far from an automatic tool. With skill, experience, and perseverance, however, it has proven in many cases to reveal almost completely what was originally written and subsequently erased. Before attempting any photographic decipherment techniques, some preliminary examination must be made. This examination plus experience with these problems generally suggest the most suitable photographic procedure.

CRIME SCENE DOCUMENTATION

The photographs taken at a crime scene are critical to an investigation. The purpose of crime scene photographs is to give a documented record of the scene as it is observed. There is a special skill and technique to crime scene photography. Therefore it takes training and practice for the photographer to be proficient in the task. First become familiar with the camera equipment – the time to become familiar with a particular piece of equipment is not during a major case when it is taken out of the vehicle. The investigator should begin taking photographs of the scene as soon as possible after arriving on the scene. This will assure that the scene is depicted as it is observed in its original uninterrupted state. Nothing should be touched, moved, or initiated into the scene until it has been thoroughly photographed and documented.

In crime scene photography there are three general positions or views, which are necessary. Those views consist of overall photographs showing the entire scene, midrange photographs showing a relationship of the items, and close-up photographs of the items of evidence themselves.

Do I use a film or digital camera? The answer to this question is both. All major crime scenes should be documented on film. Digital is used for back-up and timeliness. These images can be viewed immediately and transmitted if necessary for others to view if they are not at the scene.

Why film? A good example: a piece of evidence is not noticed until you are viewing the pictures days later. The crime scene has already been disturbed; the only documentation you have of the evidence is the pictures and it is a very small portion of the image that requires enlargement for identification. Film has such high resolution that this is not a problem. Enlarge the image in the darkroom or scan it in on a high-resolution film scanner and you make the identification of the evidence. This type of enlargement would not be possible with most digital cameras.

Use digital cameras for back-up of the film and also for the close-ups of the evidence. The resolution of the good digital cameras is quite good for most medium and close-up shots.

Each crime scene has unique characteristics and the type of photographs needed will be determined at the scene by the investigator familiar with the crime. Below are some general recommendations for covering specific scenes.

Homicide photographs (e.g., inside a residence)

- Exterior of the building
- Evidence outside the building
- Entrance into the scene
- Room in which the body was found
- Adjoining rooms, hallways, stairwells
- Body from five angles
- Close-up of body wounds
- Weapons
- Trace evidence
- Signs of activity prior to the homicide
- Evidence of a struggle
- View from positions witnesses had at time of the crime
- Suicide, other dead body calls: if there is any doubt, photograph the scene as a homicide.

Burglary photographs (residential or commercial burglaries)

- Exterior of the building
- Point of entry
- Entrance into the scene
- Interior views
- Area from which valuable articles were removed
- Damage to locks, safe, doors, toolmarks
- Articles or tools left at the scene by the suspect
- Trace evidence
- Other physical evidence
- Assaults, injuries

Face of victim

- Bruises
- Bite marks
- Orientation shot
- Close-up at 90° angle to avoid distortion
- Ruler in same plane as bite mark
- Focus carefully
- Bracket exposures

Fingerprints on scene

- Establish the location of the latent print
- Close-up to show detail
- A 1:1 camera or device must be used, or a scale must be included in the photograph on the same plane as the latent
- Photograph with the film plane parallel to the latent surface
- Get as much depth of field as possible, especially for curved surfaces
- Available light exposures of latents with normal contrast can be metered using a gray card
- Bracketing may reveal more detail in "low contrast" latents
- Underexposing the film will separate the steps on the white end of the grayscale
- Overexposure will separate the steps on the black end of the grayscale

BLOODSTAIN PHOTOGRAPHY

Many crime scene technicians are using video to document bloodstains, but whether a video camera is available or not, it is absolutely essential that still photographs be taken to document the crime scene and any associated blood evidence. If a video camera is available, then still photography will be the second step in recording the crime scene. Photographs can demonstrate the same type of things that the videotape does, but crime scene photographs can also be used to record close-up details, record objects at any scaled size, and record objects at actual size. These measurements and recordings are much more difficult to achieve with videotape.

Overall, medium range, and close-up photographs should be taken of pertinent bloodstains. Scaled photographs (with a ruler next to the evidence) must also be taken of items in cases where relative size is significant or where direct (one-to-one) comparisons will be made, such as with bloody shoeprints, fingerprints, high-velocity blood spatter patterns, etc. A good technique for recording a large area of blood spatter on a light-colored wall is to measure and record the heights of some of the individual blood spatters. The overall pattern on the wall including a yardstick as a scale is then photographed on slide film. After the slide is developed, it can be projected onto a blank wall or onto the actual wall many years after the original incident. By using a yardstick, the original blood spatters can be viewed at their actual size and placed in their original positions. Measurements and projections can then be made to determine the spatters' points of origin.

SHOE IMPRESSIONS AND TIRE TRACKS

For the purpose of this section, impressions will be defined as both two- and three-dimensional images of footwear which are found at a crime scene. For example, a two-dimensional image would be that of dusty footwear found on a piece of glass in a door at the point of forced entry to a burglary. A three-dimensional image would be a footwear impression recovered in snow.

A good-quality digital camera will work quite well shooting impressions. The camera should have adjustable f/stops (aperture settings) and adjustable shutter speeds. A normal-angle lens would be the lens of choice for footwear photography, as a wide-angle lens would tend to distort the image being recorded. Even though many footwear impressions may be photographed in daylight, a flash attachment is necessary. Make sure the flash has a feature allowing it to be held "off camera," as using a flash attachment in footwear photography is always done in this manner. A remote shutter cord and tripod should also be considered as standard equipment, as footwear photography is sometimes done in low-light situations requiring slow shutter speeds. Never attempt to handhold a camera at a shutter speed slower than 1/60th second. Finally, remember to always have a scale available and in every footwear photo you take. The reason is obvious: the purpose of taking a photograph of a footprint is to record it for possible future comparison with a suspect shoe. In order for the technician to make a print to scale from the image you provide, a scale or ruler will have to be in the image next to the impression. The authors' preferred scale is a 6-inch plastic ruler in an 18% gray color. These are available from any crime scene product supply company.

The two-dimensional print often is the most difficult to recover and document through photography. The first step to consider in taking a photograph of such an impression is to take it from the same side on which the impression was made. You would be surprised – many photographs made of impressions, after close examination, are discovered to be reversed.

Three-dimensional prints offer their own difficulty factors, but many of the same techniques are used as with the two-dimensional print. Take your time – remember, this may be the only chance you get to photograph what may be a highly important part of the on-scene investigation. Check the lighting surrounding the scene. If outdoors on a sunny day, a three-dimensional print in a soft medium such as sand or snow will probably be easy. But what about daylight shining through the glass on which the dusty print is located? Look at the lighting closely before beginning to shoot.

The first thing to consider is to fill the frame in the viewfinder of the camera with the print. This is especially important when shooting digital photographs; use every pixel available to you. Next, affix a scale next to the impression. Then:

- Take an orientation photograph to show where in the scene the impression is located.
- Take a close-up for detail.
- Use a scale on the same plane as the impression.
- Keep the film plane parallel to the plane of the impression.
- Block out ambient light and use a strong light source at different angles to find the light angle(s) giving the best detail in the impression, then put the electronic flash or light source at that angle for the photograph.
- Photograph tire impressions in sections showing one circumference of the tire.
- Use a tape measure for overlapping photographs.

SURVEILLANCE

The methodology of surveillance photography is a secretive and continuous (or sometimes, periodic) visual documentation of activities involving persons, places, or objects of importance to an investigation.

This type of photography is important in establishing identity and obtaining a permanent record of activities, in both criminal and civil court proceedings. Often surveillance will result in photographs of criminal activity in progress.

Additionally, insurance industry surveillance may produce photographic documentation of fraudulent activity by an individual wrongfully claiming physical disabilities.

Before you embark on any surveillance assignment, know what photographic equipment is needed and how to use it. Usually a faster ISO is desirable. ISO 400 speed has been found to be very satisfactory. This will require a high-end digital camera that allows the use of higher ISO settings.

Depending on the method of surveillance – whether stationary (remote or manned), on foot, or mobile – you must choose the equipment best suited for the occasion to be sure it is in working order and that all those involved know how to operate their assigned cameras, lenses, and accessories. This should also involve testing the camera under similar light conditions in a controlled situation. You will know what to expect from your camera only after you have used it. The chances of reshooting surveillance subjects are not very great. Once on the job, your primary concerns will be getting the necessary photographic documentation and not having your covert activities discovered by your subject.

The primary purpose of surveillance photography is to document evidence of importance to criminal or civil legal proceedings. If your covert activities infringe upon the subjects' rights to privacy, or if you do not have court authorization to conduct certain surveillance techniques, you may jeopardize the entire case because your photographic evidence may be ruled inadmissible.

CAMERAS AND ACCESSORIES FOR SURVEILLANCE PHOTOGRAPHY

Because surveillance work by its very nature is secretive, the equipment should be compact and easy to use and produce accurate photographic documentation. The camera may be as basic to use as the manual type, in which focusing, lens settings, and shutter speed are all controlled by the operator. On the other hand, because of activity, or in the case of a mobile surveillance, it may be preferable to use an autofocus, programmable camera. However, keep in mind that the more automatic the camera, the more batteries it requires and, consequently, the more things that can and will go wrong. For the most part, auxiliary lighting, electronic flash, bulbs, or strobes are out of the question. You must rely on available light from nature and, if you are fortunate, a strategically located streetlight or other form of illumination normally found in the area. Two exceptions would be the use of a light intensifier or starlight scope and infrared photographic techniques. Adapters are available to fit most cameras or video recorders with these low-light accessories.

These techniques produce acceptable photographic documentation in those scenarios where extremely low-level light sources exist (i.e., stars, moon, minimal street lighting). However, they are not without limitations. The light intensifier is not suitable for identifying color because the picture produced has a light green tint.

A rule of thumb for selecting lenses is 2 mm for every foot you are from the subject. This gives the best overall coverage and detail.

There are many accessories for aiding in the reduction of blurred photographs frequently encountered in low-light, slow-shutter-speed photography. Tripods, monopods, straps, and shutter release cords are essential for the use of long-focal-length lenses. A telephoto lens over 100 mm shot at slow shutter speeds (more than 1/30 second) can produce noticeable camera vibration. In situations where you find yourself without the aid of such devices, leaning against a building or any sturdy vertical support will help immensely to reduce movement. You may also steady your camera by lying over the hood of the car or even prone on the ground. A technique that will aid all slow-shutter, low-light photography work is learning to time your breathing with the squeeze of the shutter release. If you are in a situation where the subject matter does not allow for supporting the camera in any of the previously mentioned ways, use your elbows with controlled breathing in the following manner. Supporting the camera in both hands, place your elbows against your ribs. As you focus, regulate your breathing so that you exhale as you squeeze the shutter release slowly.

TRAFFIC ACCIDENT SCENES

Photographs taken at the scene of a traffic accident are a most critical part of the investigation. Most law enforcement investigators are familiar with the term "a true and accurate depiction of the scene." It is important that law enforcement personnel assigned the task of photographing traffic accidents understand more than just the basics of camera operations and exposure, so that photographs taken will truly represent the scene as it was at the time of the investigation. It becomes even more critical when using a digital camera.

Experience shows that the nighttime traffic accident is one of the more challenging scenes to photograph, particularly when skidmarks, and/or other items of evidence, are scattered over a great distance. Simply put, the camera does not see what you see. The adjustments you as the photographer make in such things as camera position, time of exposure, and supplemental lighting will be the deciding elements as to whether your camera allows the film to "see" as you do.

It is important that the camera kit includes a manual camera with adjustable shutter speeds and a "B" (bulb) setting. This "B" setting allows the shutter to be opened for an indefinite period of time, which allows for time exposures of as long as the photographer deems necessary. This immediately poses a problem with digital cameras, because only the high-end cameras would have this capability.

A normal lens with adjustable f/stops makes a perfect match for the manual camera. The term "depth of field" is familiar to most photographers. Simply put, depth of field is the amount of the scene (distance in front of the camera from its closest to farthest points) which is in focus. Although depth of field is something not of a major concern in nighttime photography, one should still understand that setting the lens aperture at f/2 will yield very little depth of field, and f/22 will dramatically increase the amount of depth of field.

A remote cable release is necessary so that the shutter can be "tripped" without having to touch the camera. Painting with light is done with very long exposures (sometimes as long as several minutes) so it is very important that camera movement be eliminated if possible. The camera will be mounted on a tripod during the photography and, with the cable release, will help eliminate the unwanted camera movement.

A powerful external flash will be essential in illuminating 150 feet of skidmarks at a nighttime accident. A rechargeable external battery for the flash is optional, but suggested. Painting with light involves illuminating the scene with multiple flashes on a single frame of film. Since several angles of the scene may be photographed, it would not be unheard of to illuminate a frame of film with 20 or more flashes on full power output.

DOMESTIC VIOLENCE

Documentation by photography is an important and powerful tool in the investigation of domestic violence crimes. When injuries resulting from domestic violence are promptly and adequately documented, it is possible for prosecution to occur without the victim's testimony. Often, victims of domestic violence are dependent upon the abuser for food and shelter. If the abuser is only jailed temporarily after the initial arrest, it is possible for an abuser to coerce the victim into not testifying. Therefore, the importance of documentation becomes relevant in preventing the recurring abuse of victims. The pictures can be used in the event that the victim later becomes unwilling to testify.

CAMERAS AND LENSES

To obtain adequate detail, close-up photography is essential. The digital camera incorporates an onboard LCD viewer that enables the subject to be viewed through the camera lens. This allows the photographer to document exactly what is framed by the view without "chopping off" the subject or getting out of focus.

For domestic violence photography a (macro) close-up capability is necessary. Most of the photographs taken in domestic violence photography will be taken at the minimum focusing distance to "fill the frame."

CHAPTER 8

LEGAL CONSIDERATIONS

Anyone who has seen a movie made in the past five years is well aware that it is possible to do almost anything you want to do with digital images – change which items are in the picture, change their coloration or texture, move things around, etc.; thus it is generally believed that digital images are easily manipulated. At the same time, there is a belief that film images are much less susceptible to manipulation. The facts of the matter are that the first statement is true and the second is, to a large degree, false. Furthermore, in a well-run lab, where proper procedures are followed, one can argue that the digital images are in fact more secure than the film images. In this chapter the issues will be reviewed and sample procedures will be provided.

TWO IMAGE CATEGORIES

The Scientific Working Group on Imaging Technology (SWGIT), operated under the leadership of the U.S Federal Bureau of Investigation, has explored the issues extensively and provides good guidelines for dealing with the technology. One fundamental distinction they have pointed out is that criminal justice images that might be used at trial fall into two categories. Some, in fact most, are "visually verifiable," and others are "analyzed." In the case of visually verifiable images, the witness uses the image to illustrate his or her memory of a scene. In essence, they will say that they were at the scene, saw the circumstances, describe key features and use the image to help the listener understand what they are describing. They could just as easily use a hand-drawn sketch. The burden is on the memory of the witness. When a photo is used, they will inevitably be asked, "Is this a fair and accurate representation of the original scene?" And the answer had better be, "Yes." The technology employed to produce the illustration is not really at issue. That is, unless there is a direct challenge to the testimony, in which case the image may come under much more scrutiny. But this is a rare event, and most of the time the details of the technology used are not important.

The story is very different when one presents an analyzed image. As an example, consider a latent fingerprint image – a dirty finger on a halftone newsprint photo. Since it is very hard to separate the halftone dots from the ridge or trough detail, it can be very difficult to read the print. In this case one might apply a Fourier analysis to selectively remove the background. This would be a clear case of an analyzed image. In this case the witness cannot say that he or she was there, saw the original object, and it looked like the image that resulted from the analysis. It did not look that way. In fact the processes used to enhance the image were specifically chosen to change the appearance, so as to render the fingerprint readable. In this case it can be argued that the witness is introducing scientific evidence, and as a result, it must be able to withstand a test in order to be admissible. In the United States, two such tests are used and referred to as the "Kelly–Frey," or "Daubert" tests – the choice of test varies by state. The basic concepts are quite similar, however. The key issues will be:

1. Is the science that was employed valid?
2. Was the science applied in a valid way?
3. Was the equipment (and software) that was used working properly at the time?

A corollary issue would be: Could another person skilled in the art obtain the same result following the indicated procedures? These issues provide a base of considerations one should consider in establishing Standard Operating Procedures (SOPs).

Certain image editing tools should be ruled out altogether. The cloning, or "rubber stamp" tool should never be used. Color copying should not be used. Addition of texture (except for normal amounts of edge sharpening or unsharp mask filters) should not be used. Cutting or copying from one image, or image location, and pasting in another should not be used. And so on. The key is that tools which enhance the entire image all at once are much more acceptable than those that selectively alter the image at a few, specifically selected spots.

To review, with visually verifiable images, one should limit the use of enhancement tools to those that are totally global, are similar to basic darkroom analogs of processes such as adjustment of brightness, contrast, and color. Some cropping is also probably acceptable. In these cases it is good practice to archive the original images, just in case there is a challenge. But it is probably not necessary to keep copious notes of the enhancement steps employed. In the case of analyzed images, the original image should also be archived, and detailed notes regarding the processes employed should definitely be kept. As regards tool selections, certain tools should be prohibited in a blanket way, and others might have certain restrictions applied.

NOMENCLATURE

The term "original image" is used a lot, and unfortunately its exact meaning can be confusing and in some cases, problematic. SWGIT, after much consideration, has defined a number of terms, which when used correctly, can help avoid confusion. It is best to start with an original image. This is defined as the first permanently recorded version of the image in question. It is used to make all the subsequent replicas that might be used, and because it is permanently recorded, one can rely on being able to obtain results that are comparable to previously produced versions by following the same steps. Note that in digital imaging, an image is a file of data, and not a physical object. This can cause problems of recalling the same image each and every time, and hence it is important to link an image file to a physical object. The concept behind calling the first permanently recorded version of an image the original image accomplishes this linking. It is recommended that a uniquely identifiable, write once read many times (WORM), long-lasting medium be used. An example would be a serial-numbered recordable CD. The index to the disk should be kept in a separate, secure, and durable file as well.

Another term that is important is "primary image." In many ways this is equivalent to the latent image in traditional film technology. Examples of primary images would include: (1) the image on a removable medium that was used in a digital camera, such as a compact flash card, (2) the image on a computer hard drive or removable drive that was created during a scanning process, and (3) an image transmitted from a camera in the field via wire or radio communication and recorded on a computer hard drive or other removable medium. In general, primary images are the direct result of the image capture process, and the medium on which they are recorded will be erased and reused as a practical matter. Clearly these images are quite susceptible to image tampering, and they should be used to create original images as soon as possible. This is analogous to the case with traditional film photography, where most of the special effects that might be employed would be applied to the latent image (for example using double exposure), either of the original negative or during preparation of a print or internegative. The process of making an original image from a primary image requires two additional definitions. One can "duplicate" an image or one can "copy" it.

A duplicate refers to a process in which all of the image file data are faithfully replicated from one generation to the next. Remember that with a digital image, the file is a highly and specifically structured string of numbers. When duplicating the image, the process employed must leave all of this information completely in tact. In other words the structure and contents of the second-generation file are exactly the same as the first. In this way the image itself has been perfectly preserved, and only the medium on which it is recorded has changed. An original image is a duplicate of a primary image that is written to a more permanent medium. A copy of an image is less restrictive. In a copy, the essence of the first generation is preserved, but the file is not an exact duplicate. For example, if one were to compress an image using a lossy compression routine such as JPEG, one would obtain a copy, and not a duplicate.

There is another important consideration when duplicating primary images so as to produce original images. Simply stated, one cannot edit an image that has not been opened. To edit an image one has to open it, view it (usually), apply an image-editing tool, and then save the edited version. One can use a process such as batch application of a series of image processing steps, for example by application of the actions pallette to a series of images, but these steps will not support selective editing of items seen in an image. They are applied without intelligence, unless exceptional steps are taken. What this means is that if one uses a drag-and-drop-like approach to duplicating unopened image files, one can create files of images that were preserved as original images before they were ever opened, viewed, or edited. If this is the SOP, and if personnel in the organization follow it, they can easily testify that indicated images are the original images. One can testify, under oath, that they were preserved before it was possible to edit them.

FILM-BASED IMAGES VERSUS DIGITAL IMAGES

It is commonly believed that film-based images are very secure, whereas digital images are very susceptible to tampering. When proper SOPs are employed, this is not the case. When proper SOPs are not employed, both are very susceptible. First consider the case of film negatives. These start out as 35 mm strips, which are frequently cut into strips of four negatives each. Often a lab employee uses a fine-pointed pen to put marks in the borders to simplify future use of the negatives, or to identify the strips so that they are less prone to misfiling after use. The strips are usually kept in a special filing cabinet designed to hold such strips. To alter an image, one would need to either break the chain of custody or act in violation of the SOPs, or both. But this could be done. The next step is to scan the negatives with a high-performance film scanner. These are readily available in graphics shops, advertising agencies, and movie production houses. Once digitized, the images can easily be edited and then written to a new piece of the same type of film. The resulting bogus negatives are cut after processing, markings are reapplied to the margins, and these strips are put back in the file drawer. It has been shown that when done with high-performance equipment and expert technique, even a highly trained expert cannot tell that the images have been altered. If all of the strips from the original strip are forged, it will not even be possible to detect a different batch of negative film. The only hard part of this process is gaining access to the original filmstrips and getting the revised images back in the file, or getting an employee in the department to deliberately alter evidence.

To alter digital images, the access or deliberate tampering issues are the same as before, assuming there is a strict chain of custody process in place in the agency and the preservation process described above is followed. Once the disks are found, they need to be read, and the key images altered. But now a major problem is encountered. Assuming the agency is using serial-numbered disks, as indicated above, the images cannot be put onto a disk with the correct serial number. And, since the SOP calls for keeping the index to the CDs in a separate, secure file, a problem will occur when one tries to read the images. The system will instantly note that the wrong disk is in the reader – an immediate indication that there has been tampering. Note that if properly recorded, the original CD will not accept additional information, or it will be immediately obvious that there is too much data on the disk. Again, the system will give an immediate indication that there has been an irregularity. So, with preservation on serial-numbered WORM CDs or their equivalent, it is extremely difficult to tamper with recorded images. In fact the problem is potentially much greater than it is with film-based images in today's world.

INADVERTENT IMAGE ALTERATION

Earlier, it was shown that image compression and image resampling can insert artifacts and/or drop elements out of an image. As a result it is strongly recommended that SOPs and equipment choices be made so as to minimize these problems. Also, all involved should be thoroughly trained in both the technology and the SOPs. Some tools, such as sharpening filters, can be reconfigured to give very pleasing images, but if not properly used, these can introduce lines surrounding items in the image. It is possible to make wounds seem more severe than they actually were, or make blemishes appear as bruises, or even wounds. The effect is similar to what one would obtain if a film-processing device did not provide sufficient agitation of the developer bath – the so-called Mackey Lines. It is not feasible to write the associated precautions into SOPs, but it is possible to assure that people are trained thoroughly before they work with digital images. And it is possible to assure that these issues are covered in the training program, as is done in training people to work in traditional film processing laboratories.

PROMPT ACTION AND MULTIPLE SHOTS

It is strongly urged that the SOPs require the creation of original images as soon as possible. In many instances, the photography is done long before there is a clear theory of the case. This means that in the early phases of the investigation, the photographers and crime lab personnel really know very little about the crime. As a result it is virtually impossible for them to alter the images to support a particular, contrived theory of the case. The material may not fit with facts and evidence uncovered later on. Hence it is good practice to preserve the images while the people involved can believably claim they did not know how to alter the images so as to support a theory that had yet to be devised. Photographic evidence is used to support other evidence much more than it is used as a "smoking gun," or as a type of "DNA" evidence. It is rarely the cornerstone of the case. Making all the parts and pieces fit together when some are locked down early in the investigation would call for a real stretch of the imagination.

Another practice that can be helpful, particularly when using a digital camera, is to take several photos of a particular part of the scene. Each photo in the set should be from a somewhat different angle and camera distance. It is particularly helpful if off-camera lighting is used. The result is a suite of images that cover a particular item, but juxtaposed to neighboring items from different perspectives. If one were to try to alter the item of interest, it would be necessary to alter it in all of the images from the suite in such a way as to keep the juxta-positions and perspective internally correct in each image and consistent among all those in the set. This would make it much more difficult to alter the content of the images in a way that is not detectable by an expert. The images in the set can be used when the material is first introduced, thus making it much less likely that opposing counsel will seek to challenge the photos for fear of tampering.

ERASED IMAGES

One aspect of capturing images using digital devices is the ability to erase images. Many law enforcement people, when they are trained to use silver halide technology, are told not to discard images. The frames are typically numbered, and if any are missing, there will be questions as to why. The implication is that the missing photos might have been exculpatory, and thus were purposely discarded. Prosecutors typically deal with missing images by asking the witness if any images were discarded, and if so why. The answer should be that the photos had technical problems or were worthless in some other obvious way. Examples would include: flash did not go off, the lens cap was on, the camera had not been focused, my finger was over the lens, the camera or subject moved during exposure. By addressing this up front, the appearance that something is being covered up is greatly reduced. The same is true with digital images taken with digital cameras.

Most digital cameras attach numbers to the images taken. Some set the counter to one each time a new, blank storage medium is inserted in the camera; others keep a running log for the life of the camera. In either case it is possible to deduce that some images are not being shown. In the case where the counter is reset, it is possible to have two or more photos for a single case with the same numbers – one from one card and the other from another. Again the disclosure of missing photos and duplicate numbering should be addressed up front to avoid the appearance of impropriety.

In the case of scans of films or reflection copy, there is no automatic numbering of images. It is up to the operator to name and store the various images. Hence, a scan can be deleted without a trace. In addition it is not possible to claim that scanned images were archived before they were ever viewed. Unless some batch process is used, scanned images appear on the computer screen as soon as the scanning is complete. It is important when scanning images to set the conditions for the scan correctly before the final scan is done. Typically there is a preview scan, and the operator is able to set things such as brightness, contrast, scan area, and resolution on the basis preliminary image. Sometimes there is some ability to adjust color as well. Attempts to make these kinds of adjustments after the image is scanned will usually result in the loss of some information; hence it is best to make them before the scan is made. This need mitigates against the use of batch scanning processes. As in the case of the digital images it is good practice to archive these images as soon as possible, and to address any special circumstances that may have been encountered and any controls that were applied when introducing the material. It does not serve any good purpose to have the opposing counsel draw out these issues and imply impropriety. Often, with scanning, the material that was scanned is available to be scanned again if there is a serious question of integrity of result, but this is not always the case. For example with questioned documents it is common practice to scan the pages, and then send them to the fingerprint examiners, where the page will be treated with chemicals that might destroy the integrity of the original document. This is not much different from the traditional film approach where instead of scanning, the document is photographed or simply photocopied. It is imperative that good notes are kept regarding what was done and in what order they were taken.

ADMISSIBLE OR INADMISSIBLE

The common belief is that evidence, including photos, is either admissible or not admissible. There is, however, a third option. The photos may be admitted with the proviso that material can be presented as to the credibility of the photographic material. In this way, the finder of fact (judge or jury) can decide how much weight to attach to the material. So while the photographer might be satisfied that the photos were admitted, the lawyer might be very unhappy because the opposition is able to bring in experts to challenge the reliability of the material and strength of the implication that can safely be drawn. Thus the photographer and the laboratory should take sufficient care to minimize susceptibility to such a challenge.

SPECIAL, PROTECTED FORMATS

Some device manufacturers include means to protect the integrity of images captured by their devices. Mostly these are associated with digital cameras, however. There are two types. In one case the image that comes from the camera is not viewable as originally brought into the computer. It must be "acquired" first. This process involves the calculation of the final color values for all of the pixels. The process is not reversible. That is, one can calculate the final color image, but once having done so, it is not possible to reconstruct the image data that came from the camera. To understand this, consider that one can compute the average of the numbers 9 and 11, and find it to be 10. Likewise the average of 8 and 12 is 10. Knowing that the average is 10 does not allow you to determine what the original numbers were. A similar process is involved here. The camera format image can be kept, however, even after it is acquired. If this is done, it can be acquired each and every time one wants to work with the image. As long as it is kept, it is immediately obvious that it could not have been edited, because in order to do so the image must be opened, viewed, changed and resaved. But there is no way to resave an image and get it back into camera format. As a result, all camera format images of this type are clearly unedited. But this approach is becoming less and less common. Most photographers do not want to have to deal with the acquisition process. They simply want to open their images. As a result, most of the newer cameras do not support this type of format. Most are moving towards JPEG format.

At least one camera manufacturer has created a JPEG-based process to protect the integrity of images. In their case, they compute a hash number from the image and encode it into the JPEG conversion. A hash number is simply a number that is computed according to some complex formula that uses the image subject information as input. The hash number that one obtains is directly linked to the particular content of the image. The slightest discrepancy in the content will result in a different hash number. One does not see anything in the image, but the computer sees the original number and can compare that to any image that is claimed to be the same. The important part of this process is that even the slightest change will result in a mismatch. So images cannot be color balanced, or have their brightness adjusted, or be cropped in any degree and still be matched up. So the approach is valuable, but its value is limited, and it only works with the products from the specific vendor. This means that if that vendor puts the feature in its midrange camera, but the agency wants a better grade camera for some applications, there is no help. Technology such as this for scanners would be more valuable, since it would help establish authenticity of these images that are otherwise less well protected.

BASIC ADMISSIBILITY TESTING

As was mentioned earlier, the typical test for admissibility of scientific evidence (analyzed images) starts with the question, "Is the science that was employed valid?" When a particular technology is new to usage in the courts, this issue will be tested repeatedly as different aspects are challenged, assuming the approach survives previous tests. This is usually done by having expert witnesses testify in a hearing. Those supporting the approach will be expected to demonstrate that they are indeed experts in the particular field, and they will be expected to cite the scientific literature that documents the reliability and validity of the under- lying science. There may be opposing witnesses claiming that the citations are flawed in some way, or there is contradictory literature. The lawyers will apply the scientific material to the law and argue their positions. Finally, the judge will decide whether to allow the evidence to be shown, to allow it with caveats to the finders of fact, or to rule the use of the evidence out.

The second key issue that will be debated at this hearing is whether the science (assuming it is found viable) was applied in a valid way. Did the way in which the measurements were made, or the data was analyzed, suggest that there is a significant probability of an erroneous conclusion? For example, if the case involves the use of the Fourier transform as applied to an image, there are questions regarding the area sampled. Many of the algorithms employed in fast Fourier transforms (FFTs) of images require that the area sampled is a perfect square. If it is not, the results are ambiguous, at best. Failure to sample a square would constitute invalid application of viable science. There are many other examples, and at times, the discussions can become quite technical. Yet the expert witnesses will have to be able to state their positions in a way that is mean- ingful and clear to non-scientists to win their arguments.

The third aspect of the test will focus on the equipment and software that were employed, and whether they were working properly at the time. In the field of digital imaging this would call into question the computer, its peripherals, and the software employed. Most people do not check their computer-related equipment, they simply respond to catastrophic failure when it occurs, and assume that all is well otherwise. But it is possible to have errors creep into a system, with the result that only slight errors are committed. It is also very easy to install a diagnostic package, and run it every day. This is the approach applied to other laboratory devices – the gas chromatograph spectrometers are checked regularly and often. It is also possible to test scanners by using test targets, and printers by printing the test target images to them. Note that the photo lab checks its film processors in much the same way. These are preventative steps that are easy to employ but it is not immediately obvious to most people to consider using them.

When combined, these portions of the test beg the question of reliability. Could another expert, skilled in the art, produce essentially the same result when starting with the same inputs? If opposing counsel asks for such a test, and the witness is not prepared to share the steps he or she had taken, the appearance of incompetence is created. A better strategy is to simply keep a log of the key steps taken when analyzing an image. Modern software packages have this as an automatic feature. The investigator simply has to turn the process on and save the result. Again, it is a simple precaution to put into practice. Now with a properly archived, original image, and a log of the steps taken, the resulting image or the interpretation of the findings is easy to demonstrate.

SAMPLES OF STANDARD OPERATING PROCEDURES

In response to the article on digital imaging that appeared in the February 2000 issue of *Law Enforcement Technology* magazine, many people have asked for sample SOPs. In response, Herb Blitzer, Executive Director of the Institute for Forensic Imaging, wrote the following two procedures. One is for a patrol officer who is provided with a digital camera and asked to use it to document certain situations. The second is for the photo specialist who will process the images that are turned in. These SOPs are merely samples and it is strongly recommended that each agency write their own as befitting their specific circumstances. For example, the SOPs one might develop for a more highly trained photographer would have additional provisions. To help in the development of specific SOPs, it is useful to point out many of the features that are built into the sample SOPs.

First of all, the sample SOPs (SSOPs) specify the camera to be used and the CDs to be used. These are fundamental specifications. The camera is one that has at least one million pixels. Also, it is not one that must resort to very high compression ratios to put several images on a floppy disk. Experience has shown that anything less than a one megapixel will give results of questionable quality, and in today's market, there are several reasonably priced, high-quality cameras that exceed this minimum. The CDs specified in this procedure are write once read many times (WORM) disks that have embedded serial numbers. This means that once the image information is recorded on one of these CDs and its index preserved elsewhere, one has a uniquely identifiable and virtually incorruptible record of the images. It serves a purpose similar to that served by negatives in traditional photography.

Most of the digital cameras in the price and complexity range suitable for this application compress the image files using the JPEG algorithm prior to storing them on the camera's removable storage medium. JPEG is an approach that is both variable and "lossy" (it permanently discards some of the information). One can select the level of compression to be used, and the cameras have a few settings for this. In most other cameras, the storage medium is a "flash card," which is capable of holding up to 8 or even 96 megabytes. These are rugged, easy to use, and reusable. Floppy disks, by comparison, hold only 1.44 megabytes, are less rugged, and like the flash cards are easy to use and are reusable. But the capacity issue is more important than it might seem. One factor is that a 5 × 7 inch print should be supported by a 4.5-megabyte image file. Another factor is that, in the camera, one wants to store as many images as possible. The cameras with floppy disks have a setting that will allow the storage of 20 or 24 images on a single floppy disk. If you do the arithmetic, this will require a compression ratio of about 65:1. At this level, significant amounts of information will be lost in the storage process. And, to make matters worse, the process by which information is lost is one that inserts artifacts into photos. One failure mode, for example, is that it has been seen that wounds are exaggerated by placing a dark ring just inside the periphery of the dark red portion, and a light ring is created just outside. In essence the edge of the wound is exaggerated, making it appear worse than it is. The process also alters colors, introducing blues and purples that may seem to be bruises, etc. In general large compression ratios are to be avoided. If one chooses a camera with a flash card storage device with 16, or even 8 MB instead of a 1.44 MB floppy disk, the effect is greatly reduced. So the bottom line is to avoid the floppy disk cameras. This is built into the SSOPs.

The next series of points have to do with the composition of the photos. Prosecutors have made it clear that they really would like to have shots of the setting, overall shots of the victim(s), and the alleged perpetrator as well if there is any question of a fight instead of battery. They also would like close-up shots of wounds with a measuring instrument in the photo. Finally, if there is appearance of drugs and/or alcohol, shots of the paraphernalia can be useful. So, the

SOP spells this out. Finally, the evidence ruler that is specified has patches of white, gray, and black. This can be a big help in color-balancing the photos later on. So this, too, is in the SSOPs.

The SSOPs make it clear that it is proper to discard images that are clearly of no value because of photographic technique or errors. Otherwise, keep all images. This avoids wasting disk space and does not weaken the case. The officer should be sure that the prosecutor trying the case is aware that some images were discarded, but it is OK to do so for the right reasons.

Because digital cameras tend to have poor image quality in the brightest portions of the scene being recorded, the SSOP advises the officer to weaken the flash during close-ups. This is a simple process and results in much better photos.

Compared to 35 mm film, digital cameras have lower quality images in areas such as resolution, dynamic range, and color fidelity. Accordingly, they should not be used as the primary photographic device in recording major crime sites. This is also addressed in the SSOPs. Also, in this regard, the SSOP for the photo specialist is restricted to photos from patrol officers. In the case of crime scene or crime lab photos taken by a more advanced photographer, or scanned from films, many additional tools can be used to extract information from the photos, which is way beyond the scope of the current application.

Once the photos are taken the SSOPs indicate that the images may be reviewed on the camera's viewer, but may not be viewed using any device that allows for editing of the images. Strict adherence to this rule makes it impossible for the officer to alter the images in a meaningful way. *In order to change an image, one must view the image and have access to tools that can make changes.* The camera allows viewing but does not support editing. Viewing the images on a laptop computer in the car would allow both viewing and modification and should not be done in this application. In special cases, where immediacy is important at the scene, one may open the images, but only duplicates of the images on the flash card should be viewed. The flash card files should not be opened or changed in any way.

The SSOPs have fairly standard provisions for turning over evidence. This is to ensure that should it become necessary to perform scientific or mathematical analyses of images, the evidence will not be easily subjected to doubt.

Once the images are in the laboratory, a process is spelled out which retains a set of unopened images. The operator will be able to swear that he or she never opened those files! Since one cannot edit images that are not visible, even if the tools are available, the integrity of the unopened images cannot have been breached. Making this the SOP, a strictly adhered-to process renders the testimony much more believable. This is followed by the prompt recording of the sets of images onto a serial-numbered, unalterable CD as soon as possible.

This, too, makes image manipulation much less likely. One cannot make an image show particular things if one does not know what those things are. Archiving the images before the case is well understood supports this, which, in turn, supports a position of not altering the images.

A minor point, but one that could get important should a case turn out to be more complicated than first anticipated, is that proper nomenclature should be used. When the camera's button is pushed, an electronic file is created inside the camera that is not visible and will soon be discarded. This file will be replicated a few times – that is, exact replicas of the data file are produced – in order to ultimately make an archive version and a visible, print version. But none of these files will be kept except for the archived replica. Accordingly, these temporary intermediates are all called "primary images." They are preliminary steps towards the creation of an archived "original image." This is slightly different from the way in which the Federal Rules of Evidence treat traditional silver halide photography today, but it is expected that the old rules will be updated to deal with the features of the new technology in the near future. Meanwhile, it is important to keep the nomenclature consistent. This SOP does this in how images at various stages are named.

In the SSOPs many of the specific operational steps are omitted. It is assumed that each individual utilizing the SOP has already had training on how to use the equipment and software involved. So there is no need to clutter up the SOP with a number of tedious, repetitive steps.

In general, there are several complex technical issues that one can encounter in using digital imaging technology. However, it is a relatively straightforward process to make the use of digital images amenable to use in law enforcement. Rather than just hope that each officer will sort all the complexities out correctly, agencies are advised to write SOPs that eliminate the guesswork. The SSOPs anticipate most of these issues and deal with them in such a way as to minimize opportunity for evidentiary challenge. These SSOPs are fully compliant with the guidelines issued by the Scientific Working Group on Imaging Technology, operating under the aegis of the Federal Bureau of Investigation.

STANDARD OPERATING PROCEDURE 1

Function Covered: Photography of incidents by a patrol officer.

Purpose: When a patrol officer is called to a scene, or comes across an apparent incident this procedure should guide the process of documentation of the situation using a digital camera.

Equipment and Materials: Each officer is issued a Kodak DCS 215 digital camera for use in such situations. In addition, there is a kit that includes a 6-inch evidence ruler with white, gray, and black patches, a battery charger, three sets of nickel-metal hydride batteries, two 16 megabyte flash cards, a line connect power source, and a download cable. Cardboard evidence envelopes are available at each district office, and it is expected that each officer will keep a few of these while on patrol.

Procedures: When covering incidents such as burglaries, domestic battery, non-injury or minor injury (the injured person does not want medical treatment) accidents, the officer is expected to determine if the investigation will be aided by photo documentation. When in doubt, take pictures, and more photos are better than fewer.

Camera settings: The camera should be set to "automatic" for exposure, flash, and focus. The viewer should be in the "on" position. The storage control should be set to "best quality." The camera viewer can be used to look at images, but the images should not be opened in a computer or any other device that allows image editing.

At the scene: Photos should include general scene shots to show what the situation looked like at the time. If there are broken windows, furniture, decorations, etc. that are indicators of violent action, care should be taken to show these situations. If there are injuries of any sort, close-up (about 10–12 inches away) photos of these should be taken with an evidence ruler in the picture. When taking close-up shots, the officer should cover part of the flash window with a finger to subdue the exposure. In addition to close-ups, broader shots of any injured people should be taken, including full-length shots. Head-and-shoulders photos of all apparent victims at a scene should be taken as well. If there are weapons or alcohol containers, or drug-related paraphernalia at the scene, these should be photographed as well. When taking photos of paraphernalia, wounds, or other items where it may later be important to know the size of the subject of the photo, two pictures should be taken: first take one of the item in its original condition, then take a second photo with an evidence ruler lying close to the subject.

If more than 10 shots are taken, the officer should write a listing of what pictures were taken and kept. He or she will need to be prepared to describe the scene and use the photos to illustrate the verbal testimony. Any photos that are photographically faulty should be discarded at the scene. These include problems such as: out of focus, flash failure, blocked lens, inadvertent push of the exposure button, blurring due to subject or camera movement, etc.

After the on-scene photography: After completing on-scene duties, the officer should place the flash card(s) used at the scene in a cardboard evidence envelope, fill out the case identification information on the envelope label, and seal the envelope. The sealed envelope should be turned in to the photo specialist at the district office in person. The photo specialist will log in the envelope, indicating the case information, the officer, and the time and date of the transfer, and both will sign that the envelope was sealed at the time of transfer. The photo specialist will then follow the Crime Scene Photo Processing Procedure to prepare the visible images and archive the records.

Prior to investigations: The officer will keep one set of batteries charging while on duty, and swap the duty batteries with those in the charger before going on call the next time. An additional set of charged batteries should be carried when on duty. This is in case power in excess of that deliverable by one set of batteries is required. Prior to leaving, the officer will check to see that the batteries are fully charged. This is to assure that the batteries in the camera are always ready for use when the officer is on duty. When working with the camera the power line connector should be used if convenient.

Calibration: While no calibration is needed, the officer is expected to check the camera on a regular basis and make sure that it is functioning normally. If there are any indications of a malfunction, the unit should be swapped with another by contacting the photo specialist.

Calculation: Not applicable.

Limitations: The digital cameras are issued to patrol officers primarily for use at scenes where photos will help the investigation and possible prosecution, and an evidence technician will probably not be called to the scene. If it is used at major crime scenes, it will serve as a back-up to the work done by the evidence technician.

Safety: There are no safety issues beyond normal precautions used when operating electrical devices. Batteries should not be heated, and if they get hot while in use or charging, they should be removed from the device. Batteries should be turned in to the photo specialist when no longer functional in exchange for new ones. They should not be discarded in any other way except in emergency.

References: In support of this procedure or use of digital cameras, one might want to consult the following publications:

- Manufacturers' literature and operating manuals, retained by the photo specialists.
- Guidelines of the Scientific Working Committee on Imaging Technology, available though the FBI.
- "Digital Imaging" by Herb Blitzer, *Law Enforcement Technology* magazine, February 2000.
- Website for the Scientific Working Groups is at www.for-swg.org. Many SWGIT materials can be found at that location.

STANDARD OPERATING PROCEDURE 2

Function Covered: Processing of images submitted by a patrol officer by the photo specialist.

Purpose: Patrol officers are instructed to take digital photos under certain circumstances and to bring the images to the Photo Specialist on flash cards. This procedure covers the processing of those images.

Equipment and Materials: A photo department image processing station, complete with peripherals and software, serial-numbered, write once read many times (WORM) compact disks (CDs), nickel-metal hydride batteries, 16 megabyte flash cards, Zip cartridges, photo-quality inkjet printer paper, ink cartridges, cardboard evidence envelopes with appropriate labels, and an evidence transfer log book.

Procedures: Each officer is instructed to bring in the flash card, with images from each incident investigated and documented with their digital camera. The flash card should be sealed in a cardboard evidence envelope with a completed case information label on the outside. The officer bringing in the envelope and the photo specialist (or deputy) receiving the envelope should enter the case information in the transfer log book, indicating the case descriptor numbers and the date and time of the transfer. Also, the envelope should be opened and the number of disks in the envelope entered. Both parties then sign the entry. Then the officer is given a new flash card(s) and new envelope. If new batteries are also required, the old ones should be swapped for new, but initially charged ones. The disk(s), stored in the envelope should be stored in the locked holding box until it is time to process the images.

When it is time to process the images, the photo specialist should remove the disks from the holding box and insert the card into the computer. Using the appropriate file management tools, two case folders should be created and labeled with the case number. One of these folders will be on the computer's hard drive and the other will be on a password-protected Zip cartridge. Then two folders should be created and inserted into each case folder – one labeled "Raw Images," and the other labeled "Processed Images" (Figure 8.1). At this point, the right-click copy tool should be used to duplicate the contents of the flash card into both of the Raw Images folders. These folders should never be opened until after the images have been permanently duplicated later on. The contents should then be duplicated a second time but placed in the Processed Images folder on the computer's hard drive. At this point some of the photos in the Processed Images folder can be opened to assure that the transfers did, indeed, occur. If all is OK, the Zip cartridge should be removed from the computer and secured. Once this is done, the "primary images" on the flash card should be erased using the disk reformatting tool. This disk is now ready to be reused by the next officer. If there are multiple disks in the case, this series of steps should be repeated until all of the images have been duplicated to the appropriate folders on both drives. The images stored in the "Raw Images" folder are now known as the "primary images."

The images in the "Processed Images" folder should be opened and *Figure 8.1*

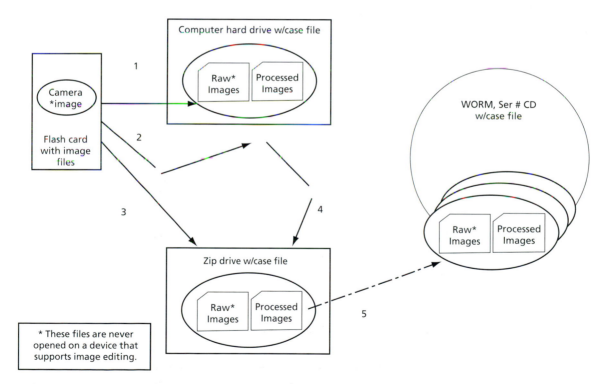

1, 2, & 3: Duplicate image files from the flash card without opening them.
4: After processing the images for printing, save processed image folder to the secure Zip drive.
5: When two secure Zip drives are full, duplicate the contents, unopened, to a serial-numbered WORM CD.

examined. All those needing adjustments for color balance, brightness, or contrast should be adjusted using the corresponding software tools. If the image needs curve shape adjustment, this should be applied as well. No further adjustments should be made. The images should then be printed, one copy of each on photo-quality paper, and another on plain bond paper. The photo-quality image should be viewed by the photographer/officer at a convenient time and then sent to the prosecutor in an internal transfer envelope with appropriate seal and case information. The plain paper copy should be filed in the secure photo office file. Images should be printed six to an $8^{1}/_{2} \times 11$ inch page. (If the prosecutor wants additional images in other sizes, they will request them afterwards.)

As soon as the images are printed, the "Processed Image" files should be duplicated onto the password-protected Zip drive, and the cartridge placed in the locked holding box. Any information regarding the images or their processing that is unusual should be noted in a text file kept in the case folder. Likewise the information that was on the label on the cardboard envelope should be entered into a text file and inserted in the case folder as well. Finally, the date and time of the processing and the name of the person processing the images should be placed in a text file and stored in the case folder.

When two password-protected Zip drives are filled, the contents should be transferred to a CD. To do this, the contents of the two drives are duplicated to a folder, and the contents of the folder are then duplicated to a CD. The cases that have been recorded on the particular CD should be listed along with the CD serial number. This record is stored in the secure photo lab file. An entry is also made into the logbook to indicate which CD (by serial number) holds the contents of the disks that were originally brought in by the officer/photographer. The images in the Raw Images folder on the CD are now known as the "Original Images." The operator should now open a few of the images on the CD to assure that the duplications occurred and the files are indeed readable. It is only now that the Raw Images folders (as recorded on the CD – the Raw Images folder on the Zip drive is never opened unless there is an emergency) can be opened and the images inside can be viewed. Once it is ascertained that the images on the CD are in fact there and readable, the Zip drive should be reformatted and it is now available for reuse. The images in the Raw Image folders are "Original" images, and all of the "Primary" images have been erased.

Calibration: The image processing system should be checked by its diagnostic each and every working day and the results filed in the secure photo laboratory office file.

Calculation: Not applicable.

Limitations: The patrol officer digital cameras are issued primarily for use at scenes where photos will help the investigation and possible prosecution and an evidence technician will probably not be called to the scene. If it is used at major crime scenes, it will serve as a back-up to the work done by the evidence technician.

Safety: There are no safety issues beyond normal precautions used when operating electrical devices. Batteries should not be heated, and if they get hot while in use or charging, they should be removed from the device. Batteries should be turned over to the district waste disposal officer when no longer viable. They should not be discarded in any other way except in emergency.

References: In support of this procedure or use of digital cameras, one might want to consult the following publications:

- Manufacturers' literature and operating manuals, retained by the photo specialists.
- Guidelines of the Scientific Working Committee on Imaging Technology, available though the FBI.
- "Digital Imaging" by Herb Blitzer, *Law Enforcement Technology* magazine, February 2000.
- Website for the Scientific Working Groups is at www.for-swg.org. Many SWGIT materials can be found at that location.

CHAPTER 9

IMAGES IN THE COURTROOM

GENERAL PRINCIPLES

At a trial, witnesses are required to swear that they are telling the "truth, the whole truth, and nothing but the truth," or words to this effect. Or, in the Biblical sense, they are not "bearing false witness." This applies to everything that they say as well as to all exhibits that they present – photos included. At a trial there is always a finder of fact. This might be a person (the judge at a bench trial) or a group of people (a jury or a judicial panel) charged with judging the veracity and implications of the material presented and deciding on guilt or innocence. One can expand this requirement a bit and address the more basic issues that underlie the oath, namely that:

- The photos are faithful, that is, neither erroneous nor misleading
- The depiction is understandable
- There is a clear implication of fairness
- The conclusions that are supported are consistent with a larger argument

FAITHFUL PHOTOS

Much of this book has been dedicated to describing tools and techniques that minimize the chance of inadvertent or purposeful distortion of images. When testifying, one is obligated to present photos that do not distort what the finders of fact would have seen if they had been at the scene. This is directly the case in terms of visually verifiable images. In the case of analyzed images, there should be testimony to support and explain any scientific or mathematical tools used to derive the images being shown. In a number of instances there may be a special hearing before the images are allowed at trial to assure that:

- The science that was employed is valid and generally accepted by the appropriate scientific community. This is usually accomplished by engaging an expert in that field to explain the science and demonstrate its applicability and the support for it through refereed journal publications. There may well be other experts called to either support the view or oppose it.

- The science was applied in a valid way in the particular instance. That is, none of the fundamental premises of the science was violated, for example: the software is such that it does not create errors, the various steps were taken in an appropriate sequence, and in general, the procedures followed are consistent with the requirements of the science.
- All equipment (and software) employed was working correctly at the time that the analysis was done. This implies that there are operational diagnostics performed and, where necessary, devices are properly calibrated.
- The images were never available to those with a reason and means to perform alterations. The normal chain of custody procedures should be followed.

All of these combine to make it particularly valuable that an agency establish standard operating procedures that are both legally and technically sound, and that all personnel learn these and follow them in all the work that they do. In the past, with conventional silver halide photography, images were generally used as visually verifiable representations. Accordingly the weight of veracity is predominantly on the witness's description of what he or she saw at the scene. In this situation, much of the requirement for SOPs is eliminated, and many organizations have taken up a very flexible approach. But today, digital technology can be employed to any image, no matter its source. Once an image has been converted to digital form, analyses can be performed much later on in an investigation and if the precursor images were not handled according to a careful procedure, the analysis and all that stems from it might be in jeopardy. "The fruit of the poisoned tree is also poisoned." The SOPs need not be overly cumbersome, but some advanced thought should be given to the procedures that might be used, and the operators should be careful to avoid certain problem approaches.

As with any photographic technique it is possible to adjust colors, brightness, and cropping so as to yield a photo that implies something other than what one would conclude if at the original scene. So, while there may not be any tampering with the items in the image, things can be rendered in a "new light" or some items can be cropped out of an image so as to provide a misleading context. Again, SOPs should be established so as to make it very unlikely that allegations such as these could be made and substantiated enough so as to produce doubt. Many of these techniques are already well known to photographers and darkroom technicians.

To assist agencies with the preparation of SOPs, the Scientific Working

Group on Imaging Technology (SWGIT) has prepared guidelines that cover these issues. SWGIT operates under the sponsorship of the U.S. Federal Bureau of Investigation (FBI). The group comprises experts in various aspects of criminal investigations at the national, regional, and municipal levels, as well as lawyers and scientists with expertise in images and image processing.

UNDERSTANDABLE DEPICTIONS

When showing a photo of a car at an accident, it is usually fairly clear to all looking at the image as to the condition of the car and where it was relative to other objects. This is even more self-evident if a series of photos is shown, each with a different perspective or at a different distance. However, even this simple situation can become confused if some care is not taken to orient the viewer. Consider that when goal-scoring sports events such as basketball or football are televised, the cameras are always on the same side of the field. There may be some shots from overhead, and some from the ends of the field, but almost never are there cameras on both sides. This is because the audience can easily become confused as to which team is headed in which direction. The same principle applies in showing images to finders of fact – especially if they are laymen. In addition, as the objects being depicted become less and less commonplace, even more care must be taken. Most people do not know what a hair looks like through a microscope. Nor are they familiar with the interpretation of fingerprints or barrel markings on a bullet, etc. In all such cases it is a good idea to show photos that make the relationships clear and the depictions understandable. With the advent of projection technology it is very easy to show several photos or diagrams to help the finder of fact understand the contents of an evidentiary image.

In some cases, neither the methodology nor the contents are commonplace. For example, photos taking advantage of fluorescent objects need some explanation. Likewise if one were to use some mathematical formulae to bring out otherwise difficult-to-see contents, it is a good idea to show a series of images to explain the means employed to create the evidentiary image. This makes it more understandable and more believable. Attempts to skip the explanation can result in disaster in some cases. If one tries to avoid sharing the means used to enhance an image, he or she is in danger of being accused of not telling "the whole truth." And even if this is done *post hoc*, it leaves the appearance of some sort of attempt to cover something up. If the process was important enough to the investigator to actually be employed, then why is it not important to share that approach with the finder of fact? It is not a good idea to have such a question even come up.

Many of the items and situations that are shown in photos are not consistent with what most people see in their daily lives. And, often the techniques used to render these items in such a way that experts can draw conclusions is even more esoteric. There is often a tendency to want to tell all of the technical steps that were taken and explain them in language rich in jargon. Instead the witness should search for and develop metaphors that relate to everyday life. For example, it is true that the Fourier analysis used to clear up fingerprints involves breaking the signal down into a sum of sine and cosine waves, and then using a graphical interface to eliminate unwanted frequencies. But this description is not very useful to the lay audience. A more commonplace metaphor would be the channel selection knob of a television set. This is attached to internal devices, which select one particular channel and reject all others. The Fourier analysis is used to accomplish a very similar task and does it in a similar way. Using the metaphor in conjunction with a series of photos at varying steps through the process can help make it all understandable to a non-scientist audience.

FAIRNESS

Finders of fact are more willing to believe an argument if it appears that the other side had every opportunity to study the premises of the argument and challenge them as necessary to uncover the truth. The discovery process, which occurs before trial, is intended to assure that this opportunity occurs and so it follows strict, due process of law. But the finder of fact does not normally see this process, since some of the evidence may be ruled inadmissible. If one side seems to have a technology advantage at trial, then it may well seem that the other side could not utilize or did not have access to such technology, and the appearance of fairness will dissolve. So if one side in a trial is using digital projection, and other advanced technology, it is wise for the other side to have access to it as well – and to be sure that the finder of fact knows this even if the opposition chooses not to utilize it. Many courts are aware of this issue and try to assure that the appearance of fairness is not compromised. Utilization of technology should clearly be per agreement of the court and if necessary, the court should provide assurances to the finder of fact as to the availability of technology to both sides.

CONSISTENT CONCLUSIONS

In the field of scientific research the conventional wisdom is that new studies should be comprise about 80% confirmation of current knowledge and 20% new learning. The same can be said of a photo being shown at a trial. To a large degree it should confirm and support material already put forward, but it should add incremental detail as well. It should be an additional strand in a multistranded cable, and not a single thread. An example of this is the traditional process for documenting a crime scene. There are initial shots showing the premises, and this is followed by a series of images showing what it looked like to someone approaching, entering, and walking though the scene. Along the way, photos of individual items of particular significance are shown. It has long been that this is the process the photographer should follow, even though only a few of the images were actually shown at trial. Now, with projection technology, it is possible to show a much larger number of images without it becoming overbearing or overly time consuming. In the proper order and with the proper description, a much clearer understanding of the crime scene can be conveyed. Along the way, the ties to the other elements of the case can be made so as to build credibility. And, at the key points, the particular new details can be inserted so as to add the new information in comprehensible context.

TWO TYPES OF PRESENTATIONS

There are two common situations in which material is introduced at a trial. First of all, there is the uninterrupted dissertation mode, and secondly, there is question–response mode. Opening arguments, closing arguments, and explanation of technical material by an expert witness under direct examination are typical of uninterrupted dissertation. The speaker is given the floor to present material in a continuous flow in which he or she can control content and order of presentation. In the question–response mode the speaker is asked a specific question and only allowed to answer that question at the time. In this mode, the questioner, and not the speaker, controls the content and order of presentation. Question–response mode typifies cross-examinations and sometimes redirects.

Clearly in the first mode, the speaker has every expectation of being able to prepare material in advance and then work through this in a linear mode. This is typical of presentations one might make at a conference or seminar. Accordingly, the sort of software often used in such applications is a good choice. In the question–response mode, this is not the case. The flow is not at all linear. There will be a need for random access to a wide range of materials and exhibits, and the selection of the next item is very likely to be based upon the response to the current question. The typical presentation software package will not be a good choice for this situation. There are specialized software packages for this situation, and it is wise to use these for the application. Most of these also have a so-called "player mode," which allows the queuing of materials for a linear presentation. These also tend to have software tools needed to prepare slides with both text and graphics.

PRETRIAL AGREEMENTS

Most courts require a discovery process by which the two sides become aware of what will be argued by the prosecutor, which witnesses might be called, and what items (including images) will be introduced during the trial. Some jurisdictions require full disclosure, which means that the prosecutor will show all evidentiary materials to the defense to assure that no exculpatory material is being withheld. To avoid problems later on, each of the images that might be used should be uniquely identified and made available to both sides in the same format. With images, especially electronic ones, there is the possibility of confusion as to which is which, particularly when there are a few that closely resemble each other. Attaching proper identification will prevent the possibility of showing the wrong image, or having the two sides looking at separate images but thinking they are the same. Also, with agreement beforehand as to the technology to be employed, there is no possibility for an image to be rendered in one way by one lawyer's equipment and another way by the other's. The use of serial-numbered WORM CDs will help assure that there are no errors at trial. All of the images will stay the same, with the same identifications, and the system used to display them will remain constant. If there are certain items that were originally proposed, but it is decided that they will not be shown, they can be eliminated before the final disks are made. Thus there is preliminary agreement that anything on the disk is potentially admissible at trial.

PHYSICAL ARRANGEMENTS

Whenever the courtroom set-up is not fixed, some attention should be paid to the physical arrangement of the displays. In many courts, it is considered proper protocol to follow a special sequence when showing an item for the first time. Assuming the prosecutor is introducing an image, he or she will first show the image to the judge, then to the defense counsel, and finally to the finder of fact. This precludes showing anything to the finder of fact that they are not supposed to see according to prior agreements. To follow this protocol, each lawyer will need a means to find the correct exhibit with no one else seeing the process, then, to have it appear before the judge, then the opposing counsel, and finally to all others in the room.

This full protocol is not always followed, however, and the issue should be discussed prior to trial so that proper arrangements can be made. In more and more cases, the courts are simply allowing the lawyer to introduce previously discovered images to all parties at the same time. This allows the use of a single, large display screen and helps everyone focus on the same image at the same time. It can help eliminate confusion when a physical pointer is used.

In some jury trial courtrooms, a set of smaller displays is placed in front of the members of the jury so that they can look at the screens in groups of two or three. This has the effect of having these people looking at the little screens and not seeing the bigger picture of events taking part live in the courtroom. A better approach is a large screen with a data projector that can provide large, bright images in a normally lit courtroom. Separate monitors can be provided for the judge and opposing counsel as required. That screen should be placed so that the jury can see both the screen and the witness in the same view. In this way the live action is associated with the presentation and the body language of the speaker can be integrated with what he or she is saying and showing in a more normal, human interaction.

To fill a reasonably large screen in a normally illuminated (but not brightly lit) courtroom will require a data projector with at least 1000 lumens of output – 2000 will be much better. A higher output machine will have a fan and therefore make some noise when running, but the noise will not be a problem in the typically sized courtroom. Most projectors have a standby mode, which leaves the electronics active, but turns off the lamp. It is a good idea to go into standby mode if the projector will not be used for the next half hour or more. But by leaving the system in standby, it can be brought back into operation much more quickly and with no fussing with controls and adjustments. The lamps in these devices require a few moments to turn on and stabilize. This is the case whether going from standby to active or going from off to active. Also, the lamps run at very high temperature and tend to heat up all of the projector parts near them when they are running. They also have rather limited lifetimes (on the order of 700 hours), are hard to find, and are quite expensive. This all means that care must be taken to not have a lamp go out during a trial. The machine will have to cool down for a few hours before anyone can reach in and change the lamp. And, if a spare is not on hand, it may be a problem to find one without significant delay.

Most projectors have sound reproduction capability. All are able to project video, either from a VCR or from the computer. But the ones with sound capability can play back the soundtrack as well. Many of the presentation software programs have the ability to add animation to slides, and to accompany the motion with sound effects. The use of these features in courtroom presentations should be greatly limited. The idea is to use the images as illustrations to the spoken testimony, and the text to amplify certain key points. Turning the presentation into something resembling a TV commercial for an e-business is inappropriate, to say the least. One should not test the limits of assuring that the probative value outweighs the inflammatory impact.

Video reconstructions are becoming easier and easier to produce. These are where motion images are shown to indicate what the action would have been like. There are problems with these, however. Consider an accident scene. It is possible to have clear data as to where the wrecked cars ended up. It is also reasonably clear as to where the cars were just prior to impact, although there is some conjecture involved. As to rates of speed at that point in time, there is usually a large degree of conjecture. And, finally, most of what is shown just prior to impact and during and just after impact is mostly conjecture. There are usually very few facts available to support the motion that might appear in a video reconstruction.

PREPARATION OF PRESENTATIONS

While all experienced trial lawyers are also experienced in preparing and delivering presentations, most are not experienced in the use of data projectors, and computers with presentation software. To deliver a voice-only presentation when one has the power to present a much stronger message via the technology would be a waste and would seem inappropriate. Juries today expect accomplished presentations, and to not deliver them is to lose some credibility in their hearts. Some of the major issues are explored in the following paragraphs.

DIGNITY OF THE PRESENTATION

It must be remembered that material presented in a trial is expected to carry a level of gravity that is consistent with the occasion – society is trying to determine if a particular individual has committed a certain crime and should be punished accordingly. This is not a lighthearted matter, and it would be a mistake to not maintain the level of decorum that this requires. Accordingly, one should avoid the use of overly fancy graphics and overly active animation effects. In addition, many courts require that the screens actually shown do not include computer toolbars and the like. The screen should be filled only by the material to be presented. Backgrounds should be relatively plain so as to not seem inflammatory relative to the evidentiary nature of what is being seen. Written words should be common and in good taste. In general, the prosecutor should be seen as trying to show commission of a crime by an individual, and should not appear to be vindictive or on a witch hunt. Likewise the defense should be trying to show that there is legitimate doubt that the prosecutor's argument is completely true, and should not appear to be seeking sympathy, bending words, hiding behind technicalities, or evasive. In both cases, the approach should be to provide logic that is straightforward and to do so with a presentation that is clear, to the point, and not reliant on pure emotions. This is not to say that the presentations should be boring, lacking in creativity, or lacking in expertise in the tools at one's disposal when using presentation technology.

ORDER OF PRESENTATION AND EASY-TO-FOLLOW LOGIC

There are many ways in which one can order the material in a presentation, but one of these stands out in being easy to follow and grasp. The basic structure is: tell them what you are going to tell them, tell them, and tell them what you have told them. This process starts with introspection. The author must sort out inputs from outputs and then sort outputs from outcomes. Clearly the inputs will comprise the evidentiary materials themselves, the testimony that went with them, and processes that were used to extract specific inferences from them. The inferences are the outputs. So, for example, photos of fingerprints and fingerprint analysis tools are inputs. The inference drawn is that either a pair of prints were made by the same person, or not. This is an output. Outcomes are larger inferences. So, for example, the print may make it clear that a certain person was at a certain place at a certain time, or held a specific item, or had access to a specific document. If the location was the crime scene, then the outcome is strength for the assertion that the person had the opportunity to commit the crime. If the item was a weapon, the outcome is that the person had the means to commit the crime. And, if the document contains information that would have enraged the person, then the person had motive. Added together, the inputs (the fingerprints and analytical tools employed, along with testimony given) allow one to assert that the accused had all of the classic antecedents to prove guilt, and that therefore he or she can reasonably be found "guilty." This argument has a few levels of inputs, outputs, and outcomes.

The presenter should start out by saying what the evidence will show. The intent is to craft "the song they will go out singing." In show business, this means that the people leaving the theater will be singing the songs that made the biggest impact on them. In presentations, it means that if you stopped one of the audience members in the hallway as they left the presentation and asked them what the presentation was all about, they would recite a few key outcomes. Experience indicates that there must be at least three and not more than five key outcomes. If there are less than three, the presentation will be seen to have been lacking in content. At the other end, people will seldom remember more than five key points from a single hearing of the presentation. So then, the way to start the presentation is to say that "We will show that A, B, C, D, and E occurred." Points A to E should be stated as simple declarative sentences. Next, the presentation should expound on the five outcomes chosen. Only material that supports one of these outcomes should be shown. This material, element by element, should be shown in a way that implies it is valid, relevant, and supportive of the indicated outcome. Once all of this is done, the three to five major outcomes should be reviewed so as to set the facts as the pillars of the conclusions drawn. The result is a series of pyramids built in layers. At the top, the overall conclusion is seen to comprise three to five smaller pyramids (outcomes). These, in turn, comprise smaller pyramids, which are built of facts, that is, inputs, be they images, testimony, analyses, or physical objects.

MIXING WORDS AND IMAGES

Software packages used for presentations all provide the ability to put small amounts of text on a screen along with images. Indeed, most have the tools needed to insert text documents from either a word processor file or from a scanned hardcopy document. They will also support use of short video files, sound files, photos, graphs, and datasheets. Such products have the means to apply animation effects as well. This is where items on the screen can be made to appear after the initial slide is shown, or items can be made to move, or vanish, and so on. Also, there can be a graceful transition between slides. Clearly these tools can be used to the point where the presentation looks more like an on-line advertisement for another website – too flashy to be taken seriously. At the same time they can also be used to make legitimate points clearer to the audience and should indeed be used.

For example, a slide can have a title appear when the slide is first seen. The result is to make it clear what will be addressed in the next few minutes. Then a first bullet point appears, starting the process of making the first point. Bullets are usually sentence fragments that have only a few words, but are made into sentence equivalents in the audience members' heads by context and words spoken by the presenter. Very soon thereafter, an image can be brought in to start to build the context. This process is often referred to as a "selective reveal" and it is used to prevent the audience from rushing ahead and reading other points and losing focus and concentration on the point at hand.

It is generally poor form for the speaker to read the words on the slide. The audience can read these much faster than the speaker can say them, and besides the redundancy will become boring. Instead the speaker should assume that the words have been read and proceed directly to amplify the text presented and point out details in the images that might accompany the text. The images become linked to the text and the thoughts and become means to evoke those thoughts later on. Later on, a simple flashing of the image is often all that is needed to bring the whole issue back to mind. This is the same idea as is behind the use of corporate or product logos. And on it goes, one set of text at a time, or combination of text with images, backed up by the spoken words of the speaker. The impact is much stronger than either the spoken words alone, or those words with the written text.

In addition to the use of simple bullets for text, most leading presentation software packages provide tools for creation of graphs, organization charts, and general-purpose box-and-arrow charts. These tools are rather easy to use and can be used along with photos, text, and animation tools. There is also great flexibility in the use of colors, borders, and other elements that can help get points across – either directly, or more subtly through associations.

JUXTAPOSITIONS

Just as bulletized text, along with spoken words can be amplified by good graphics, so can graphics items be amplified by other graphics items. The relative locations can have dramatic impact. At a recent trial in Indiana in the USA, a prosecutor's opening statement built a pyramid of photos. A nurse was on trial for killing seven of his patients. The first slide showed a small family photo of an elderly woman in the lower left hand corner. The speaker introduced her and gave her background and why she was in the hospital. The next thing the audience saw was the appearance on the screen of a second similar type of photo of an elderly gentleman, just to the right of the first photo. Again the speaker introduced this second person. The process continued until there were four such photos across the slide. Then a fifth image appeared, above the first two on the left, bottom row. A similar set of introductions continued until there were seven people in all, arranged in a sort of pyramid. At this point all of the victims had been introduced. Then, abruptly, a black-and-white, dour mug shot appeared atop the pyramid, looking down on the victims. This photo was larger and the implication of power was palpable. Moreover, the family-like photos of the victims were in stark contrast to that of the defendant. This done, the prosecutor went on to explain what the State would prove over the next several days of the trial. In the closing argument, the defendant was shown in the center of a slide, and then one by one photos recalling the key parts of the case presented at trial appeared in a box-like pattern. The defendant wound up "surrounded" by the evidence. Then, after a review of the case, the prosecutor went back to the slide shown at the opening, with the seven victims under the control of the defendant. And, to clarify the nature of the crime, and add drama in the process, the victims' images dissolved out of the slide one by one. The overall effect was dramatic, clear, and based upon the evidence presented at trial.

Using photos in conjunction with line drawings, other photos, highlighting, and other graphic techniques can make the argument in a very compelling way, even with no words on the page. Just carefully timed words spoken by the lawyer or expert witness. These tools should be used in prudent, creative, and well-developed presentations to take full advantage of the technology.

UNRESOLVED ISSUES

The use of digital technology in courtrooms is relatively new, and while there is a well-established beginning, there are certainly lessons to be learned. In this book, processes for capturing, storing, enhancing, and presenting images have been shown. These concepts are based on considerations from many aspects by experts from several fields of specialization (through work with many jurisdictions and the work of SWGIT). They should go a long way to support procedures that will hold up well in actual court proceedings in many jurisdictions.

RECORD

One of the issues yet to be satisfactorily resolved is that of creating a valid transcript of the trial. Traditionally, a court reporter captures all the words that are spoken and transcribes this into a written document that can serve to support post-trial issues such as appeals. In addition, all materials that were placed in evidence are preserved as part of the record. In many cases this will be in the form of paper photographs of physical items and written reports by experts. But what about a presentation that was supported by the use of a graphics presentation package – how should this be preserved? Some have resorted to creating a video record of the entire proceedings. This can be cross-referenced to the transcript to show what really was seen, when it was seen, and what was being said at the time. Unfortunately the quality of such tapes is really not up to the task and there is not a good way to preserve the record for more than a few years. In courts with more advanced display systems, a digital record can be created. This is more amenable to preservation, but since most of the application software employed will probably be obsolete in just a few years and the equipment used will be unavailable in just a few years more, it may not be possible to play back the record in a time frame consistent with the rules of court in the jurisdiction. It is not impossible to develop solutions to these potential problems, but they are not in widespread use today.

PROPER DECORUM

At the one extreme one can envision a terribly boring presentation with no clever use of graphics presentation tools, and at the other extreme a circus-like presentation that is devoid of character and therefore credibility. A question to be resolved is how far one can go to enliven a presentation before the boundaries of proper behavior are violated.

FIDELITY OF PRESENTATION

Most presentation graphics programs do not have good image editing capabilities. So images have to be prepared for presentation prior to inserting them into a presentation. One key factor is that the images will be slow in getting to the screen if the image files are too large. Typically an image of 300 kilobytes or so will be just fine when shown in such a presentation and it will not be a problem getting to the screen quickly. However, if the image was prepared by one person with the understanding that the full image will be used, and the presentation is put together by a second person who wants to crop out a portion of the image and then use it to fill a large area of the screen, problems can arise. Presentation software packages automatically resample the image for presentation at the screen resolution indicated and for the size of the image selected. This can lead to degradation of the image. Under certain situations it may not be possible to demonstrate the intended point very well. This will be particularly true if there are specific items, such as thin straight lines or text in the images. In addition, if certain image compression algorithms, such as JPEG, are used at too high a compression ratio, these effects can be compounded. There are ways to deal with these issues, but each particular organization has to work out the processes it will follow so as to avoid problems.

More and more police agencies are adding the ability to provide digital images to trial lawyers; however, there is not much in the way of standards. It is important that the police agencies in an area coordinate their choices with the prosecutors and defense lawyers in that same area. This is simple to do in advance and will avoid compatibility problems and thereby greatly reduce the costs associated with revisiting past work.

PRESENTATION SOFTWARE FOR RANDOM ACCESS

Because presentation software is widely used by the general business public, it is widely available, there are many courses one can take to learn how to use it, it accepts a wide range of file formats for insertions, and it is reasonably priced. Also, it tends to be updated frequently. Software intended for use in the courtroom is targeted at a much smaller market, and accordingly it does not have as many training vehicles, it will probably not be updated as frequently as general public packages, and over time this will result in some limitations on compatibility with file formats.

When preparing a presentation using a general public presentation package, one can be rather certain that the court will have the software already installed, and all that is needed is to bring the presentation file. On the other hand, if one is using a less widely distributed software package, this assumption cannot be made. One must either make advance arrangements for it to be installed, or one must bring it along and load it.

It is hard to predict the high-tech industry very well, but it seems likely that the vendors of popular presentation software packages might well add abilities to provide random access to a set of visual assets and thus mimic the main attractions of specialty software. But only time will tell. Today, to be fully prepared for opening statements, direct examination, cross-examination, and closing arguments, one needs to have software intended for trials.

ADDITIVE/PRIMARY COLOR
Red, green and blue, are the three colors used to create all other colors when direct or transmitted light is used (such as on a computer monitor). They are called additive primaries, because when pure red, green and blue are superimposed on one another, they create white.

ADDRESSABLE RESOLUTION
The maximum number of pixels a graphic expansion board is capable of manipulating. This may, or may not, be the same amount the graphics board or monitor is capable of displaying.

ALPHA CHANNEL
An eight-bit channel reserved by some image-processing applications for masking or additional color information.

ALPHANUMERIC
Both numbers and letters.

ANALOG
A form of measurement or representation in which an indicator is varied continuously, often to reflect ongoing changes in the phenomenon being measured or represented. A wave of water is in analog form.

ANAMORPHIC SIZING
Unequal scale change in the horizontal and vertical direction of a scanner. This enables the scanner to adjust the ratio in the horizontal and vertical direction.

ANTI-ALIASING
A filtering technique to smooth jagged edges on raster image screen computer graphics. A technique for merging object-oriented art into bitmaps.

APPLICATION PROGRAMS
Programs supplied by computer manufacturers and software vendors that make the computer perform useful functions, such as word processing or Adobe PhotoShop.

ARCHIVE
To copy programs, pictures or data to an auxiliary storage medium such as removable disks or magnetic tape, for long-term retention.

ARRAY
An ordered collection of elements of the same type. Represented by a single line of sensors in a CCD chip are called a linear array. A digital image is stored as a 2-dimensional data array containing pixels addressable by x, y (or row, column) coordinates.

ARTIFACT
A visible indication (defect) in an image caused by limitations in the reproduction process (hardware or software).

ASCII
American Standard Codes for Information Interchange (pronounced "ask-key"), a computer code that offers compatibility between computers and programs.

BACKGROUND MODE
IBM's and compatibles and Macintosh can perform some functions, while

being used for other purposes. For example, an operator may be able to manipulate one digital image, while simultaneously producing a print from another image in the background mode.

BANDING
A visible stair-stepping of shades in a gradient.

BAUD RATE
The speed at which computer data is transmitted and is equal to bits per second. Today, modems operate at 28.8 baud and higher.

BEZIER CURVES
In object-oriented programs, a curve whose shape is defined by anchor points along its arc.

BINARY SYSTEM
A number system with a base of two; therefore, each digit has only two possible states (0 or 1). Binary numbers are the basis for internal computer language.

BIOS
Basic input/output system – the computer part that manages communications between the computer and peripherals, such as the keyboard, monitor and I/O ports.

BIT
One binary digit (0 or 1), the smallest amount of information a computer can handle.

BIT-MAPPED GRAPHIC
A graphic image formed by a pattern of pixels (screen dots) and limited in resolution to the maximum screen resolution of the device being used. Paint programs usually produce bit-mapped graphics.

BITMAP-TYPE IMAGE
A single-channel image with 1-bit of color information per pixel, also known as a bitmapped image. The only colors displayed in a bitmapped type image are black and white.

BRIGHTNESS
One of three dimensions of color; the other two are hue and saturation. The term is used to describe differences in the intensity of light reflected from or transmitted through an image independent of its hue and saturation.

BUBBLE MEMORY
Data storage using magnetized dots. Offers high capacity, but is very expensive.

BUFFER
Temporary electronic storage space for computer memory, often used to hold graphic or text information.

BUG
A programming error that can cause unexpected results.

BULL'S EYES (REGISTRATION MARKS)
Marks that appear on a printed image, generally for CMYK color separations, to help you align the various printed plates.

BUNDLING
Combines two or more functions into one expansion board or computer peripheral device.

BUS
Path for transmitting data. Expressed as an x-bit bus, for example 16-bit bus. The larger the bus, the faster the data is transferred. AT class machines have a 16-bit bus, and 486/pentium class machines have a 32-bit bus.

BYTE
Eight binary digits or bits. A byte can take on values from 0 (00000000) to 256 (11111111) and store one ASCII character.

CALIBRATION BARS
The printed 11-step gray-scale wedge that appears on printed output. When you print a CMYK color separation, this step wedge appears only on the back plate. On a color image, this refers to the color swatches printed at the sides of the image.

CAP
Computer-Aided Photography or Computer Assisted Photography, such as the use of internal camera positioning sensors, with a computer interface, to calculate and position tilt, shift and swing adjustments on a studio camera, as with the Sinar E view camera.

CCD
Charge Coupled Device – a computer chip with light receptors and a key element in digital cameras and scanners.

CD-ROM
Compact Disk "Read Only Memory" – an information storage system using compact disks to hold up to 670 MB of read-only computer data. Playback is through a CD-ROM drive unit. One disk can hold the equivalent of many volumes of conventionally printed text or graphics, but information cannot be changed or added.

CD-I
Compact Disk "Interactive" - A system using CD-ROM disks with the added ability of responding to computer operators actions – often used in training or education.

CHANNEL
Analogous to a plate in the printing process, a channel is the foundation of an image. Some image types have only one channel, while other types have several channels. An image can have up to 16 channels.

CHIP
An integrated computer microprocessor circuit etched onto a silicon chip, which determines its capabilities.

CLIPBOARD
A temporary storage area in memory where text and/or graphics are stored as you copy or move them. Caution, because the clipboard can hold only one unit of cut or copied information at a time, moving text with the clipboard can be risky.

CLIPPING
Process of setting graphics display boundaries. The window defines the clipping volume, near and far clipping planes, and projectors of the corners of the window. Data on the planes forming the edges are considered to be within the volume.

CMYK
The subtractive colors/four colors used in printing: cyan, magenta, yellow and black. Black is usually added to enhance color and to print a true black.

CMYK IMAGE
A four-channel image containing a cyan, magenta, yellow and black channel. A CMYK image is generally used to print a color separation.

COLOR CORRECTION
The changing of the colors of pixels in an image, including adjusting brightness, contrast, mid-level grays, hue and saturation to achieve optimum printed results.

COLOR PICKER
Utility for specifying colors on the monitor.

COLOR SEPARATION
The division of an image into component colors for printing. Each color separation is a piece of negative or positive film.

COLOR SPACE
Three-dimensional model (or representation of a 3D model) used to organize colors to show progressions of hue, lightness, and saturation. Device-indepen-

dent color spaces are based on international standards (CIE).

COMPOSITE MONITOR
A low cost computer color monitor, which receives its color and brightness information in one signal. (See RGB video)

COMPRESSION RATIO
A processing technique for increasing a storage system's ability to hold information by applying mathematical algorithms to the digital data. Compression ratios of 10 times are not unusual.

CONTINUOUS-TONE IMAGE
A color or grayscale image containing gradient tones from black to white.

CONTRAST
The tonal gradation between the highlights, midtones and shadows in an image.

CROP
To select part of an image and discard the unselected areas.

DECOMPRESSION
Process used to change compressed files back into their original form.

DENSITY
The ability of an object to stop or absorb light. The less light that is reflected or transmitted by an object, the higher the density. Also, in electronic imaging, the number of characters that can be stored in a given physical area. As density increases, data storage increases.

DENSITY RANGE
The range from the smallest highlight dot the press can print to the largest shadow dot it can print.

DIALOG BOX
In a graphical user interface, an on-screen message box that conveys or requests information from the user.

DIGITAL
Information or graphic data that has been translated into a discrete numerical value and can, therefore, be manipulated and reproduced without loss of quality.

DIGITAL IMAGE
An array of pixels, each of which has its own color. When viewed on a monitor or paper, the digital image appears like a photograph.

DIGITAL PHOTOGRAPHY
Producing or reproducing an electronic image represented by a series of numbers, which can be manipulated by computer and then reconstructed as a photographic image.

DIGITIZE
To transform a continuous tone image into computer readable data using a device called a scanner, or the process of converting an analog image to digital information usable by a computer.

DIRECTORY
The computer's filing system.

DITHERING
In color or gray scale printing and displays, the mingling of dots of several colors to produce what appears to be a new color. With dithering, 256 colors can be combined to produce what appears to be a continuously variable color palette, but at the cost of sacrificing resolution.

DOT GAIN
Effect in the printing process, which causes dots making up a photo to print larger and darker than they were intended. Dot gains of 30 percent are common in the mid-tones of an image and lesser amounts in the highlights and shadows.

DOTS PER INCH (dpi)
A measure of screen and printer resolution that counts the dots per linear inch the device can produce.

DOUBLE-CLICK

To click the mouse button twice in rapid succession. In many programs, double clicking extends the action that results from single clicking. Double clicking is also used to initiate an action in a file list. For example, double-clicking a file name selects and opens the file.

DOUBLE DENSITY

A widely used recording technique that packs twice as much data on a floppy or hard disk as the earlier single-density standard.

DOWNLOADING

The reception and storage of a program or data file from a distant computer through data communication links.

DRAG

To move the mouse pointer, while holding down the mouse buttons.

DRIVER

Software the computer uses to make another device perform a prescribed function.

EDIT MODE

A program mode that makes correcting text and data easier.

ELLIPTICAL DOT

Type of halftone screen dot with an elliptical, rather than circular shape, which sometimes produces better tonal gradations.

EPS

Encapsulated PostScript, this format carries a pict preview and is the only format that supports saving line screen data and transfer functions. In bitmapped mode, it also supports transparent whites.

EXPORT

To output data in a form another program can read.

FAST COMPRESSION

Compression program built into a board, within the computer.

FEATHER EDGE

The area along the border of a selection, that is partially affected by changes to the selection.

FILE NAME

A name assigned to a file so the operating system can find the file. You assign file names when the files are created. Every file must have a unique name in accordance with the system you are using (IBM or MAC).

FILE SERVER

Computer attached to the network that sends messages or files from one network user to another, stores a backup copy of software programs, stories or art files and can send files to different printers on the network.

FILL

To paint a selected area with a gray shade, a color or a pattern.

FLATBED SCANNER

A scanner with a non-moving surface, upon which artwork is placed. Scanning CCD sensors pass across the material, producing a digital image.

FLOATING SELECTION

A selection that has been moved or pasted on an image. It floats above the pixels in the underlying image until it is deselected.

FLOPPY DISK

A removable and widely used secondary storage medium that uses a magnetically sensitive flexible disk enclosed in a plastic envelope or case.

FRAME GRABBER

A computer expansion board with the ability to digitize a single video or TV image for use in digital photography.

GAMMA
A measure of contrast that affects the mid-level grays (midtones) of an image.

GAMMA CORRECTION
Compressing or expanding the ranges of dark or light shades in an image.

GAMUT
The ranges of hues that can be reproduced from a given technology, process, and set of colorants in all combinations.

GCR
Acronym for Gray Component Replacement. A technique for reducing the amount of cyan, magenta and yellow in an area and replacing with an appropriate level of black.

GIGABYTE
A unit of measure of stored data corresponding to one billion bytes of information.

GRABBER HAND
In graphics programs, an on screen image of a hand you can position with the mouse to move selected units of text or graphics from place-to-place on the screen.

GRADATION
Smooth transition between black and white, one color and another or color and the lack of it.

GRADIENT FILL
A fill that displays a gradual transition from the foreground to the background color. Gradient fills are made with the blend tool.

GRAPHICS
In personal computing, the creation, modification and printing of computer-generated graphic images. Two basic types are object-oriented (draw programs) and bit-mapped (paint programs) graphics.

GRAPHICS TABLET
A graphics input device that enables you to draw with an electronic pen on a electronically sensitive table. The pen's movements are relayed to the screen.

GRAY-SCALE IMAGE
A single-channel image consisting of up to 256 levels of gray, with 8 bits of color information per pixel. Since each dot making up a pixel can be adjusted by the eight bits of information, this works out to a math formula of two to the power of eight or 256.

HALFTONE
An image created with a pattern of data for different sizes used to simulate a continuous tone photograph, either in color or black and white. The halftone screen converts continuous-tone copy to line copies (discrete dots of varying sizes and shapes) for printing on a press.

HANDLE
In an object-oriented graphics program, the small black squares that surround a selected object, enabling you to drag, size or scale the image, commonly referred to as transformations.

HARDWARE
The electronic components, boards, peripherals and equipment that make up your computer system – distinguished from the programs (software) that tell these components what to do.

HIGH DENSITY
A storage technique for secondary storage media such as floppy disks. This technique requires the use of extremely fine-grained magnetic particles. High-density disks are more expensive to manufacture than double density disks. High-density disks, however, can store one megabyte or more of information on one 5 1/4-inch or 3 1/2-inch disk.

HIGHLIGHT
The lightest part of an image, repre-

sented in a halftone by the smallest dots or the absence of dots.

HISTOGRAM
A graphic representation of the number of pixels with given color values. A histogram shows the breakdown of colors in an image.

HSL IMAGE
An RGB image that is displayed in three channels – hue, saturation and luminance – with only one channel displayed at once.

HSI
Acronym for Hue, Saturation and Intensity.

HSV
Acronym for Hue, Saturation and Value.

HUE
The main attribute of a color that distinguishes it from other colors.

HUNG SYSTEM
A computer that has experienced a system failure sufficiently grave to prevent further processing, even though the cursor may still be blinking on the screen. The only option in most cases is to restart the system, which entails loss of any unsaved work.

ICON
In a graphical user interface, an on-screen symbol that represents a program file, data file or some other computer entity or function.

IMAGE COMPRESSION
The use of a compression technique to reduce the size of a graphics file (which can consume inordinate amounts of space).

IMAGESETTER
Device used to output a computer image or composition at high resolution onto photographic paper.

IMPORT
To load a file created by one program into a different program.

INDEXED COLOR IMAGE
A single-channel image, with 8-bits of color information per pixel. The index is a color lookup table containing up to 256 colors.

INTERFACE
An electronic circuit that governs the connection between two hardware devices and helps them exchange data reliably.

JPEG
Acronym for Joint Photographic Experts Group. Group proposing a standard for file compression.

K
Notation for black when dealing with color images.

LAN
Acronym for Local Area Network. Hardware and software used to connect a number of computers in a small specific area together.

LPI
Acronym for lines per inch. Measure of the frequency of halftone screen.

LUMINANCE
Lightness; the highest of the individual RGB values, plus the lowest of the individual RGB values; divided by two; a component of an HSL image.

MAGNETIC DISK
In secondary storage, a random access storage medium that is the most popular method for storing and retrieving computer programs and data files. In personal computing, common magnetic disks include 3 1/2-inch floppy disks, zip drives, jazz drives and hard disks of various sizes.

MAGNETIC OPTICAL DRIVE
Hard disk storage device that uses a laser to store and read data on a magnetically sensitive disk. This disk is removable in a cartridge case.

MASK
Inactive area or a bitmapped image that will not respond to changes.

MENU BAR
In the industry standard and graphical user interfaces, a bar stretching across the top of the screen (or the top of the window) that contains the names of pull-down menus.

MENU-DRIVEN PROGRAM
A program that provides you with menus for choosing program options so you do not need to memorize commands.

MERGE
To combine two or more groups of data into one larger arranged set of data, such as two pictures.

MIDTONE
Tonal value of dot, located about halfway between the highlight value and the shadow value.

MODEM
A device that converts digital signals generated by the computer's serial port to the modulated, analog signals required for transmission over a telephone line and transforms incoming analog signals to their digital equivalents.

MOIRÉ PATTERN
An undesirable pattern in color printing, resulting from incorrect screen angles of overprinting halftones. Moiré patterns can be minimized with the use of proper screen angles.

MONITOR CALIBRATION
Process of correcting the color rendition of the monitor to match the printed output.

MOUSE
An input device, equipped with one or more control buttons, housed in a palm-sized case and designed to roll about on the table next to the keyboard. As the mouse moves, its circuits relay signals that moves a pointer on the screen.

NODE
A terminal or station that, along with others, is connected to one common computer or is part of a computer network.

NOISE
In an image, pixels with randomly distributed color values.

OK BUTTON
In the industry standard and graphical user interfaces, a push button in a dialog box the user can activate to confirm the current dialogue box settings and execute the command.

OPERATING SYSTEM
A master control program for a computer that manages the computer's internal function and provides you with a means to control the computer's operations.

OPTICAL DISK
A secondary storage medium for computers in which you store information of extremely high density on a disk in the form of tiny pits, the presence or absence of which corresponds to a bit of information read by a tightly focused laser beam. One optical disk can store as much as one billion bytes of data.

PAGINATION
Electronic manipulation of text and graphics so an entire page is printed at one time.

PAINT FILE FORMAT
A bit-mapped graphics file format found in programs such as Macpaint and PC Paintbrush.

PALETTE
The total number of colors a graphic expansion board is capable of manufacturing. Or, the total number of colors a graphic expansion board is capable of displaying on a monitor at one time. Also refers to boxes that open inside an application program to perform actions on the image.

PANTONE MATCHING SYSTEM
A standard color-selection system for professional color printing supported by high-end graphic programs, such as Adobe PhotoShop.

PCX
A file format for graphics created by the Paintbrush programs.

PERIPHERAL
A device, such as a printer or disk drive, connected to and controlled by a computer, but external to the computer's central processing unit.

PHOTO CD
A digital imaging system from Eastman Kodak that stores over 100 photographic color images on one CD disk. With correct software, such as Adobe PhotoShop, the digital images can be transferred from a CD-ROM drive to a computer hard drive and be manipulated. Photo CDs are usually produced from traditional 35mm negatives and slides.

PIXEL
The smallest element (a picture element) a device can display on-screen and out of which the displayed image is constructed.

PIXEL INTERPOLATION
A technique for increasing the size of a graphic file by creating pixels to produce pseudo-detail in an image.

PIXEL REPLICATION
A technique for increasing the size of a graphics file by duplicating existing pixels a set number of times.

PLUG-IN MODULE
Software usually developed by a third-party vendor in conjunction with Adobe Systems that lets you use a function that is not available in the standard Adobe PhotoShop application.

POINTING DEVICE
An input device such as a mouse, trackball or stylus graphics tablet used to display a pointer on screen.

PRINTER DRIVER
A file that contains information a program needs to print your work with a given brand and model of printer.

PROCESS COLOR
The four color pigments – cyan, magenta, yellow and black – used in color printing.

PROMPT
A symbol or phrase that appears on-screen informing you that the computer is ready to accept input.

PULL-DOWN MENU
A method of providing a command menu that appears on-screen only after you click the menu's name.

RADIO BUTTON
In a graphical user interface, the round option buttons that appear in dialog boxes. Unlike check boxes, radio buttons are mutually exclusive; you can pick only one of the radio button options.

RAM
Random Access Memory – the computer's working memory, usually expressed in megabytes.

RASTERIZATION
Process of converting mathematical and digital information into a series of dots by an imagesetter for production of negative or positive film.

REPEAT RATE
The rate at which paint is deposited on an image by the painting and editing tools when the mouse is stationary.

RESAMPLE
To change the resolution of an image. Resampling down discards pixel information in an image; resampling up adds pixel information through interpolation.

RESET BUTTON
A button, usually mounted on the system unit's front panel that enables you to perform a warm boot if the system is hung so badly the reset key doesn't work.

RESOLUTION
A measurement – usually expressed in linear dots per inch (dpi) or pixels per inch (ppi), horizontally and vertically – of the sharpness of an image generated by an output device such as a monitor, scanner, camera or printer. The number of pixels per inch in an image; or the number of dots per inch by an output device. Resolution can also refer to the number of bits per pixel. Resolution is usually stated in rows and columns of pixels, such as 1,000x1,200 pixels.

RGB IMAGE
A three-channel image containing a red, green and blue channel. The additive primary colors.

RGB VIDEO
A computer monitor or system that employs separate red, green and blue signals on separate cables, resulting in higher resolution.

RIP
Acronym for Raster Image Processor. Device found inside a laser printer or laser typesetter where electronic signals are controlled. The RIP is made up of software and hardware whose job it is to interpret the signal generated and sent by the computer and then form a pattern of dots so the printer's recorder can draw it on paper using a laser beam.

ROTATION TOOL
In graphics or desktop publishing programs, an on-screen command option, represented by a icon, that the user can use to rotate images or type from its normal horizontal position.

SCALING
In presentation graphics, the adjustment of the y-axis chosen by the program so that differences in the data are highlighted. "Changes the size of the image."

SCANNER
A peripheral device that digitizes artwork or photographs and stores the image as a file that can be merged with text in many word processing and page layout programs.

SCREEN CAPTURE
The storage of a screen display as a text or graphics file on disk.

SCREEN FREQUENCY
The density of dots on the halftone screen commonly measured in lines per inch. Also known as screen ruling.

SCROLL
To move the window horizontally or vertically so its position over the image changes.

SCSI
Small Computer System Interface, pronounced "scuzzy." – an interface used by personal computers to communicate with peripheral equipment.

SERIAL PORT
A port that synchronizes and makes asynchronous communications between the computer and devices such as serial printers, modems and other computers easier.

SHADOW
The darkest part of an image, represented in a halftone by the largest dots.

SOFTWARE
System, utility or application programs expressed in computer-readable language.

SPACING
The distance between the pixels that are affected by each painting and editing tool.

STORAGE DEVICE
Any optical or magnetic device capable of secondary storage functions in a computer system.

SUBMENU
A set of lower level commands available when you choose a top-level command.

TAGGED IMAGE FILE FORMAT (TIFF)
A bit-mapped graphics format for digital images. TIFF simulates gray-scale shading.

THERMAL-DYE SUBLIMATION PRINTER
A printer capable of producing continuous-tone, photographic-quality color prints from digitized images. Heat transfers colors to a support paper, using a dye sublimation process. One of the most advanced units of this type is the Kodak 8650 color printer.

TOLERANCE
A parameter of the magic wand and paint bucket tools that specify the color range of the pixels to be selected.

TOOLBOX
The set of tools normally displayed to the left of the image. The toolbox is a floating palette that can be moved or hidden.

UNDO
A program command that restores the program and your data to the stage they were in just before the last command was given or the last action was initiated.

UPLOAD
To transmit a file by telecommunications to another computer user or bulletin board.

USER-FRIENDLY
A program or computer system designed so that persons who lack extensive computer experience or training can use the system without becoming confused or frustrated.

USM (Unsharp mask)
The term comes from a conventional color separation camera technique that uses a Unsharp photographic mask to increase contrast between light and dark areas of the reproduction and gives the illusion of sharpness.

VIDEO IMAGE CAPTURE BOARD
Frame digitizing board inside the computer that accepts a video signal and converts the signal into a digital image.

VIRUS
A computer program designed as a prank or as sabotage that replicates itself by attaching to other programs and carrying out unwanted or sometimes damaging operations.

WINDOW
A rectangular on-screen frame through which you can view a document or other application.

WYSIWYG
An acronym (pronounced "wizzy-wig") for "what-you-see-is-what-you-get", which refers to the visual fidelity of the computer screen display and the printed result.

INDEX

WITHDRAWN